Ocular Stem Cell Studies

Jeffrey N. Weiss

Ocular Stem Cell Studies

A Comprehensive Overview of Global Clinical Trials

Springer

Jeffrey N. Weiss
Parkland, FL, USA

ISBN 978-3-031-58556-2 ISBN 978-3-031-58557-9 (eBook)
https://doi.org/10.1007/978-3-031-58557-9

This Springer imprint is published by the registered company Springer Nature Switzerland AG
The registered company address is: Gewerbestrasse 11, 6330 Cham, Switzerland

For the patients that have allowed us the privilege to serve.

Preface

This book is a compendium of the worldwide ocular stem cell studies registered with clinicaltrials.gov. The presented information is accurate as of July 2023. I have divided the studies into multiple categories: Active/Recruiting, Active/Not-Recruiting, Completed, Unknown Status, Withdrawn, Terminated, and Suspended. Within each category, Retinal/Optic nerve studies are listed first, followed by Anterior segment studies. Regarding study location, United States studies are listed first, followed by other countries listed in alphabetical order.

I have corrected the mischaracterization of studies, to only include those that truly belong within each category, and to correct spelling and grammar, without changing the spirit or intentions of the submitters, so as to produce a consistent and easy-to-read format. The Study title is provided, as is the country of origin and the clinical trial number in order to make it easier for the reader to locate the study and obtain further information.

"Withdrawn," "Terminated," and "Suspended," studies are also provided for their learning value.

The "Analysis" chapter summarizes study results.

The "Comments" chapter discusses the various issues facing ocular stem cell researchers.

I hope that by providing this reference, the field of ocular stem cell research will be advanced.

Parkland, FL, USA

Jeffrey N. Weiss

Contents

Chapter 1
Active/Recruiting Ocular Stem Cell Studies

United States

Retina/Optic Nerve

Recruiting

Stem Cell Ophthalmology Treatment Study II (SCOTS2)

ClinicalTrials.gov ID NCT03011541

Sponsor MD Stem Cells
Information provided by MD Stem Cells (Responsible Party)
Last Update Posted 2023-06-06

Brief Summary
This study will evaluate the use of autologous bone marrow-derived stem cells (BMSCs) for the treatment of retinal and optic nerve damage or disease.

Detailed Description
Eyes with loss of vision from retinal or optic nerve conditions generally considered irreversible will be treated with a combination of injections of autologous bone marrow-derived stem cells isolated from the bone marrow using standard medical and surgical practices. Retinal conditions may include degenerative, ischemic, or physical damage (examples may include macular degeneration, hereditary retinal dystrophies such as retinitis pigmentosa, Stargardt, non-perfusion retinopathies, post retinal detachment). Optic Nerve conditions may include degenerative, ischemic, or physical damage (examples may include optic nerve damage from glaucoma, compression, ischemic optic neuropathy, optic atrophy). Injections may include retrobulbar, subtenon, intravitreal, intraocular, subretinal, and intravenous.

J. N. Weiss, *Ocular Stem Cell Studies*,
https://doi.org/10.1007/978-3-031-58557-9_1

Patients will be followed for 12 months with serial comprehensive eye examinations including relevant imaging and diagnostic ophthalmic testing.

Official Title
Bone Marrow-Derived Stem Cell Ophthalmology Treatment Study II

Conditions
Retinal Disease
Age-Related Macular Degeneration
Retinitis Pigmentosa
Stargardt Disease
14 More Conditions

Intervention/Treatment
Procedure: Arm 1

Study Start (Actual)
2016-01

Primary Completion (Estimated)
2024-07

Study Completion (Estimated)
2025-07

Enrollment (Estimated)
500

Study Type
Interventional

Phase
Not Applicable

Other Study ID Numbers
SCOTS2

Eligibility Criteria
Inclusion Criteria

- Have objective, documented damage to the retina or optic nerve unlikely to improve OR.
- Have objective, documented damage to the retina or optic nerve that is progressive AND have less than or equal to 20/30 best-corrected central visual acuity in one or both eyes AND/OR an abnormal visual field in one or both eyes.
- Be at least 3 months postsurgical treatment intended to treat any ophthalmologic disease and stable.
- If under current medical therapy (pharmacologic treatment) for a retinal or optic nerve disease be considered stable on that treatment and unlikely to have visual function improvement (for example, glaucoma with intraocular pressure stable on topical medications but visual field damage).

- Have the potential for improvement with BMSC treatment and be at minimal risk of any potential harm from the procedure.
- Be over the age of 18.
- Be medically stable and able to be medically cleared by their primary care physician or a licensed primary care practitioner for the procedure.
- Medical clearance means that in the estimation of the primary care practitioner, the patient can reasonably be expected to undergo the procedure without significant medical risk to health.

Exclusion Criteria

- Patients who are not capable of an adequate ophthalmologic examination or evaluation to document the pathology.
- Patients who are not capable or not willing to undergo follow-up eye exams with the Principal Investigator or their ophthalmologist or optometrist as outlined in the protocol.
- Patients who are not capable of providing informed consent.
- Patients who may be at significant risk to general health or to the eyes and visual function should they undergo the procedure.

Ages Eligible for Study

18 years and older (Adult, Older Adult)

Sexes Eligible for Study

All

Accepts Healthy Volunteers

No

General Publications

- Weiss JN, Levy S. The Role of Patient Funded Clinical Research in Advancing
- Medical Care. ClinicalTrials.gov Identifier: NCT01920867.
- Weiss JN, Levy S, Malkin A. Stem Cell Ophthalmology Treatment Study (SCOTS) for Retinal and Optic Nerve Disease: a Preliminary Report. Neural Regeneration Research. 10;6: 982–988. 2015.
- Weiss JN, Levy S, Benes SC. Stem Cell Ophthalmology Treatment Study (SCOTS) for Retina and Optic Nerve Disease: Case Report of Improvement in Relapsing Auto-Immune Optic Neuropathy. Neural Regeneration Research. 10;9:1507–1515, 2015.
- Weiss JN, Benes SC, Levy S. Stem Cell Ophthalmology Treatment Study (SCOTS): Improvement in Serpiginous Choroidopathy following Autologous Bone Marrow Derived Stem Cells. Neural Regeneration Research 11:1512–1516, 2016.
- Weiss JN, Benes SC, Levy S. Stem Cell Ophthalmology Treatment Study (SCOTS): Bone Marrow Derived Stem Cells in the Treatment of Leber Hereditary Optic Neuropathy. Neural Regeneration Research. 11:1685–1694, 2016.
- Weiss JN, Levy S. Neurologic Stem Cell Treatment Study (NEST) using bone marrow derived stem cells for the treatment of neurological disorders and inju-

ries: study protocol for a nonrandomized efficacy trial. Clin Transl Degener Dis. 12;1(4):176–180, 2016.
- Weiss JN, Benes SC, Levy S. Stem Cell Ophthalmology Treatment Study: bone marrow derived stem cells in the treatment of non-arteritic ischemic optic neuropathy (NAION). Stem Cell Investigation. November 2017. http://sci.amegroups.com/issue/view/632.
- Weiss JN, Levy S. Autologous bone-marrow derived stem cells in the treatment of "untreatable" optic nerve and retinal conditions. EC Ophthalmology 9.5(2018):332–336.
- Weiss JN, Levy S. Stem Cell Ophthalmology Treatment Study: bone marrow derived stem cells in the treatment of Retinitis Pigmentosa. Stem Cell Investigation. https://doi.org/10.21037/sci.2018.04.02.
- Weiss JN, Levy S. Dynamic Light Scattering Spectroscopy of the Retina – A non-invasive quantitative technique to objectively document visual improvement following ocular stem cell treatment. Stem Cell Investig. 2019;6:8 doi https://doi.org/10.21037/sci.2019.03.01.
- Weiss JN, Levy S. Stem Cell Ophthalmology Treatment Study (SCOTS): Bone marrow derived stem cells in the treatment of Usher syndrome. Stem Cell Investig. 2019;9:9. https://doi.org/10.21037/sci.2019.08.07.
- Weiss JN, Levy S. Stem Cell Ophthalmology Treatment Study (SCOTS): Autologous bone-marrow derived stem cells in the treatment of hereditary macular degeneration. EC Ophthalmology 10.7 (2019): 536–542.
- Weiss JN, Levy S. Stem Cell Ophthalmology Treatment Study (SCOTS): Bone marrow derived stem cells in the treatment of Dominant Optic Atrophy. Stem Cell Investig. 2019; https://doi.org/10.21037/sci.2019.11.01.
- Weiss JN, Levy S. Stem Cell Ophthalmology Treatment Study (SCOTS): Bone marrow derived stem cells in the treatment of Age-related Macular Degeneration. Medicines 2020, 716;1–13. https://doi.org/10.3390/medicines7040016.
- Weiss, JN, Levy, S. Stem Cell Ophthalmology Treatment Study (SCOTS): Bone Marrow-Derived Stem Cells in the Treatment of Stargardt Disease. Medicines 2021, 8, 10. https://doi.org/10.3390/medicines80200101-10.

Recruiting

Human iPSC for Repair of Vasodegenerative Vessels in Diabetic Retinopathy (iPSC)

ClinicalTrials.gov ID NCT03403699

Sponsor University of Alabama at Birmingham
Information provided by Maria Grant, University of Alabama at Birmingham (Responsible Party)
Last Update Posted 2023-07-05

Brief Summary

This study proposes to carefully examine the hypothesis that human inducible pluripotent stem cells (iPSCs) can be effectively employed as a future therapeutic option for individuals with diabetic retinopathy and macular ischemia. iPSCs will be generated from the peripheral blood cells of subjects with diabetes and age-matched controls. The human iPSC cells will be used to generate mesoderm cells for injection into the vitreous cavity of diabetic rodents and primate eyes. The ability of mesoderm cells to generate endothelial cells and pericytes in areas of degenerated capillaries will be examined. The human iPSCs will also be used to generate hematopoietic CD34 + CD45+ cells. The combination of CD34 + CD45+ cells derived from iPSCs and iPSC-derived mesoderm will be examined in combination for their potentially beneficial effect to enhance the vessel formation.

Detailed Description

Vascular complications due to diabetes mellitus (DM) are the result of sustained vascular injury with insufficient vascular repair. In chronic diabetes, vascular reparative mechanism can be lost resulting in development of microvascular complications (MVC), such as diabetic retinopathy (DR). We assessed the reparative function of progenitor cells that circulate in the peripheral blood of diabetic individuals and found that the vascular wall-derived progenitor cells, endothelial colony forming cells (ECFCs), were depleted in diabetics with MVC. Bone marrow-derived progenitor cells, CD45 + CD34+, were dysfunctional in diabetics with MVC. We found that human-inducible pluripotent stem cells (hiPSCs)-derived ECFCs displayed the ability to form functional and durable blood vessels in vivo and conferred therapeutic revascularization by connecting with and remaining integrated with host rodent vessels long term. We characterized a mesoderm subset (SSEA5-KNA+ cells) generated from hiPSCs that gives rise to ECFCs. Finally, we used hiPSCs to generate CD34 + CD45+ cells and tested the impact of coadministration of these cells with ECFCs within the vitreous. The addition of CD34 + CD45+ cells with ECFCs resulted in the enhanced survival, function, and reparative ability of the ECFCs. This beneficial effect was mediated by reducing retinal oxidative stress and inflammation.

These novel and paradigm shifting findings led us to hypothesize: the hiPSC-derived mesoderm subset (SSEA5-KNA+) can be utilized for long-term revascularization of vasodegenerative capillaries and their reparative action can be further enhanced by coinjection of CD34 + CD45+ cells that provide antioxidant and anti-inflammatory effects.

Official Title

Human iPSC for Repair of Vasodegenerative Vessels in Diabetic Retinopathy

Conditions

Diabetes Complications
Diabetic Retinopathy

Study Type

Observational

Enrollment (Estimated)
20

Intervention/Treatment
Biological: Generation of inducible pluripotent stem cells

Other Study ID Numbers
300000173

Study Start (Actual)
2018-01-11

Primary Completion (Estimated)
2024-01-31

Study Completion (estimated)
2025-01-31

Study Contact
Name: Jennifer Moorer
Phone Number: 2053258674
Email: jmoorer@uabmc.edu
United States

Birmingham, Alabama, United States, 35294
Recruiting
University of Alabama at Birmingham

Contact
Jennifer Moorer
205-325-8674 jmoorer@uabmc.edu
Principal Investigator:
Maria B Grant, MD

Eligibility Criteria
Ages Eligible for Study
21–98 years (Adult, Older Adult)
Accepts healthy volunteer

Sexes Eligible for Study
All

Sampling Method
Non-Probability Sample

Study Population
Patients who have retinal abnormalities other than diabetic retinopathy will be excluded. Patients who have history of malignant disease or hematologic disorder. We will record medications that the patient is taking at the time of biopsy. Baseline characteristics will also be recorded, including age, lipid parameters, body mass index, blood pressure, smoking history, antioxidant intake, and use of nutritional supplements.

Description

Inclusion Criteria

- Any man or woman between the ages of 21–98 years will be eligible to participate. To participate in the study as a study subject, we will require: (a) the subject must either carry the diagnosis of diabetes or be a healthy aged control and (b) the patient be willing and have the ability to cooperate with the eye exam and skin punch biopsy protocol.

Exclusion Criteria

- We will apply the following exclusion criteria: (a) evidence of ongoing acute or chronic infection (HIV, Hepatitis B or C, tuberculosis); (b) ongoing malignancy; (c) cerebral vascular accident or cerebral vascular procedure; (d) current pregnancy; (e) history of organ transplantation; (f) presence of a graft (to avoid any effect of the graft on inflammatory parameters); and (g) patients with anemia. Subjects with AMD, glaucoma, uveitis, known hereditary degenerations, or other significant ocular complications other than diabetic retinopathy will be excluded.

Design Details

Observational Model: Cohort
Time Perspective: Other
Biospecimen Retention: Samples With DNA
Biospecimen Description: Inducible pluripotent stem cells will be generated and cryopreserved.

Cohorts and Interventions

Group/cohort	Intervention/treatment
Nondiabetics Any man or woman between the ages of 21–98 years will be eligible to participate. To participate in the study as a study subject, we will require: (a) the subject must be a healthy control and (b) the subject be willing and have the ability to cooperate with the eye exam and blood draw	Biological: Generation of inducible pluripotent stem cells • Generation of inducible pluripotent stem cells from peripheral blood cells
Diabetic Any man or woman between the ages of 21–98 years will be eligible to participate. To participate in the study as a study subject, we will require: (a) carry the diagnosis of diabetes and (b) the subject be willing and have the ability to cooperate with the eye exam and blood draw	Biological: Generation of inducible pluripotent stem cells • Generation of inducible pluripotent stem cells from peripheral blood cells

Primary outcome measures

Outcome measure	Measure description	Time frame
Generating iPSCs from peripheral blood	Blood will be collected from the patient and cells will be isolated and shipped to ALSTEM for generation of iPSCs	From blood draw to 4 months
Differentiate iPSCs into CD34+ cells and mesoderm	Specific cell culture conditions will be used to differentiate the cells into these two distinct populations	4 months to 4 years

Sponsor
University of Alabama at Birmingham

Collaborators
No information provided

Investigators
- Principal Investigator: Maria B Grant, MD,1954

General Publications
No publications available

Recruiting

CNS10-NPC for the Treatment of RP

ClinicalTrials.gov ID NCT04284293

Sponsor Cedars-Sinai Medical Center
Information provided by Clive Svendsen, Cedars-Sinai Medical Center (Responsible Party)
Last Update Posted 2023-02-13

Brief Summary
The investigator is examining the safety of transplanting cells into the subretinal space of patients with Retinitis Pigmentosa (RP). The cells are called neural progenitor cells, which are a type of stem cell that can become several different types of cells in the nervous system. These cells have been derived to specifically become astrocytes, which is a type of neuronal cell. The cells are called "CNS10-NPC." The investigational treatment has been tested in animals, but it has not yet been tested in people. In this study, the investigators want to learn if CNS10-NPC cells are safe to transplant into the subretinal space of people.

Detailed Description
This study will be the first to use a human progenitor cell line to treat retinitis pigmentosa in people. This is a Phase 1/2a, single-center, open-label, safety study of two escalating doses of human neural progenitor cells (CNS10-NPC) delivered unilaterally to the subretinal space of subjects with RP.

Subjects meeting all Eligibility Criteria and providing Informed Consent will be enrolled in one of two sequential dosing groups. Subjects will be treated sequentially with a minimum of one-month interval between surgeries for the first three subjects in each dosing cohort. The remaining subjects in the cohort will be treated with a minimum interval of at least one week between surgeries.

Primary objective. To assess the safety and tolerability of two escalating doses of clinical-grade human fetal cortical-derived neural progenitor cells (CNS10-NPC) administered in the subretinal space of one eye (unilaterally) in patients with retinitis pigmentosa (RP).

Secondary objectives. Within constraints of a small first in-human study focused on safety:

1. Determine if CNS10-NPC can engraft and survive long-term in the retina of transiently immunosuppressed subjects.
2. Obtain evidence that subretinal injection of CNS10-NPC can favorably impact the progression of vision loss in subjects with moderate RP.

Official Title
Clinical Study to Assess Safety and Efficacy of Subretinal Injection of Human Neural Progenitor Cells for Treatment of Retinitis Pigmentosa

Conditions
Retinitis Pigmentosa

Study Type
Interventional

Enrollment (Estimated)
16

Intervention/Treatment
Biological: CNS10-NPC implantation

Phase
Phase 1

Other Study ID Numbers
Study 00000375

Study Start (Actual)
2021-07-22

Primary Completion (Estimated)
2024-11

Study Completion (Estimated)
2024-11

Study Contact
Name: Pablo Avalos, MD
Phone Number: 310-248-8584
Email: Pablo.Avalos@cshs.org
United States

Beverly Hills, California, United States, 90211
Recruiting
Retina-Vitreous Associates Medical Group

Contact
Janet Kurokouchi
310-289-2478 ext. 3 jkurokouchi@laretina.com
Principal Investigator:
David Liao, MD, PhD

Eligibility Criteria
Ages Eligible for Study
18 years and older (Adult, Older Adult)

Accepts Healthy Volunteer
No

Sexes Eligible for Study
All

Description

Inclusion Criteria
- To participate in this study, the subject must be 18 years of age or older and must understand and sign the protocol's informed consent.
- Participants with diagnosis of retinitis pigmentosa.
  - Clinical signs of retinitis pigmentosa
  - A history of nyctalopia
  - Retinal pigmentary changes
  - Arteriolar attenuation
  - Waxy disc pallor
  - Electrophysiologic evidence of rod dysfunction on full field electroretinography
  - Visual field constriction.

  Participants in Group 1 ($n = 6$) will have visual acuity equal to or worse than 20/200. Participants in Group 2 ($n = 10$) will have visual acuity equal to or worse than 20/80.
  Group 1 participants will have central visual field of 40 degrees diameter or less. Group 2 participants will have a measurable visual field defect.
- Participants will be medically able to undergo ophthalmic surgery.

Exclusion Criteria
- Presence of significant ocular abnormalities that would preclude the planned surgery or interfere with the interpretation of study endpoints (e.g., glaucoma, corneal, or significant lens opacities, preexisting uveitis, intraocular infection, macular edema, choroidal neovascularization). Any ocular diseases that the investigator feels would interfere with accurate ocular measurements. This would exclude potential subjects with significant cataract, corneal scars, or significant corneal irregularities such as keratoconus, previous penetrating keratoplasty, or glaucoma with central visual field.

- Any preexisting factor or history of eye disease that may predispose to an increased risk of surgical complications in the study eye (e.g., trauma, previous surgery other than uncomplicated cataract surgery, uveitis, congenital developmental or structural abnormalities).
- Concomitant systemic diseases including those in which the disease itself, or the treatment for the disease, can alter ocular function or immune status (e.g., malignancies, uncontrolled diabetes).
- Any ocular surgery or laser in either eye within 12 weeks of screening.
- Any contraindication to pupil dilatation in either eye.
- Treatment with intravitreal, subtenon, or periocular steroid within 4 months of enrollment.
- Any known allergy to any component of the delivery vehicle or diagnostic agents used during the study (e.g., dilation drops) or medications planned for use during the perioperative period including corticosteroids, tacrolimus, and mycophenolate.
- Imminently life-threatening illness.
- Abuse of alcohol or any illegal substance(s) within 12 months of the procedure.
- Laboratory test abnormalities or abnormalities in electrocardiogram or chest X-ray, which in the opinion of the Principal Investigator are clinically significant and would make the patient unable to tolerate study procedures.
- Intercurrent illness or infection 28 days prior to enrolment.
- Contraindications to use of anesthesia.
- Females of childbearing potential (e.g., those who are not surgically sterile and not at least 12-months postmenopausal) who are not willing to comply with the study's contraception requirement (women who are unwilling to use an effective form of contraception such as the contraceptive pill or intrauterine device).
- Women who are pregnant.
- Females who are nursing or who intend to nurse during the first 6 months posttreatment.
- Men or women who do not agree to use barrier or medical contraception for at least 6 months postoperatively.
- History of any investigational agent administration within 28 days prior to administration.
- Participation in a prior gene transfer therapy study or cell-based therapy.
- Enrollment in any other clinical study, for any condition, including those relating to RP throughout the duration of the study.
- Current or anticipated long-term treatment with systemic corticosteroids (for a period longer than 7 days).
- Current treatment with immunosuppressant therapies or any contraindications to use of the immunosuppressants in this protocol.
- A history of malignancy within a five-year period or a positive cancer screening test within a one-year period of the screening visit.
- Any physical or mental disability that will impair the ability of the patient to travel to and from the study center or provide informed consent/assent or effective safety assessments as specified by the protocol. Any mental or psychiatric

disorders that prevent the patient from having full authority to do so (e.g., the subject cannot have power of attorney signed over to another individual).
- Inability or unwillingness to comply with the study protocol.
- Medical history of HIV, or hepatitis A, B, or C.
- Subjects who have taken any prescription or investigational oral retinoid medication (e.g., Accutane® Soriatane®) or any medicines that may affect the macula (e.g., tamoxifen, mellaril, thorazine, plaquenil, niacin) within 6 months of enrollment.
- Allergy to Beta-Lactam antibiotics.
- The presence of cystoid macular edema.
- Glucose-6-phosphate dehydrogenase (G6PD) deficiency.

Design Details
Primary Purpose: Treatment
Allocation: Non-Randomized
Interventional Model: Sequential Assignment
Masking: None (Open Label)

Arms and Interventions

Participant group/arm	Intervention/treatment
Experimental: Group 1A Visual acuity of 20/200 or worse Single, unilateral, subretinal injection of 300,000 CNS10-NPC ($n = 3$)	Biological: CNS10-NPC implantation • All patients will receive a single, unilateral, subretinal injection of CNS10-NPC • Other Names: – Neural Progenitor Cell (NPC) – Stem Cell
Experimental: Group 1B Visual acuity of 20/200 or worse Single, unilateral, subretinal injection of 1,000,000 CNS10-NPC ($n = 3$)	Biological: CNS10-NPC implantation • All patients will receive a single, unilateral, subretinal injection of CNS10-NPC • Other Names: – Neural Progenitor Cell (NPC) – Stem Cell
Experimental: Group 2 Visual acuity between 20/80 and 20/200 Single, unilateral, subretinal injection of 1,000,000 CNS10-NPC ($n = 10$)	Biological: CNS10-NPC implantation • All patients will receive a single, unilateral, subretinal injection of CNS10-NPC • Other Names: – Neural Progenitor Cell (NPC) – Stem Cell

Primary outcome measures

Outcome measure	Measure description	Time frame
Safety as evaluated by incidence of Adverse Events (AE) and Serious Adverse Events (SAE) and their relationship to the intervention		Subjects will be followed postoperatively for 12 months
Safety, as evaluated by changes in the complete blood count		Subjects will be followed postoperatively for 12 months

Safety, as evaluated by changes in the comprehensive metabolic panel		Subjects will be followed postoperatively for 12 months
Safety, as evaluated by changes in Donor-Specific Antibodies		Subjects will be followed postoperatively for 12 months
Safety, as evaluated by changes in the urinalysis		Subjects will be followed postoperatively for 12 months
Safety, as evaluated by changes in Visual Acuity	ETDRS or FrACT in cases of very low vision	Subjects will be followed postoperatively for 12 months
Safety, as evaluated by changes in visual field	Goldman perimetry will be used for central visual fields and microperimetry will be used for peripheral visual fields	Subjects will be followed postoperatively for 12 months
Safety, as evaluated by changes in Spectral Domain Optical Coherence Tomography (SD-OCT)	Ellipsoid zone measurement through SD-OCT will be performed	Subjects will be followed postoperatively for 12 months

Secondary outcome measures

Outcome measure	Measure description	Time frame
Slit-lamp examination		Performed 7 times over 15 months
Intraocular pressure measurement		Performed 7 times over 15 months
Funduscopic examination		Performed 7 times over 15 months
Fundus photography (color, red free, or infrared)		Performed 4 times over 15 months
Fundus Autofluorescence (FAF)		Performed 4 times over 15 months
Best-Corrected Visual Acuity		Performed 7 times over 15 months
Visual Function Questionnaire-25 (VFQ-25)	The Visual Function Questionnaire-25 is graded in a scale from 0–100 where a higher number represents better function	Performed 5 times over 15 months
Spectral domain optical coherence tomography (SD-OCT)	Retinal Thickness	Performed 5 times over 15 months
Electroretinography (ERG)		Performed 5 times over 15 months
Goldmann visual field area (microperimetry)		Performed 5 times over 15 months
Change in rate of vision field loss		Through study completion, ~15 months

Sponsor
Cedars-Sinai Medical Center

Collaborators
- California Institute for Regenerative Medicine (CIRM)

Investigators
- Principal Investigator: David Liao, MD, PhD, Retina-Vitreous Associates Medical Group

General Publications
No publications available

Recruiting

Safety and Tolerability of RPE Stem Cell-derived RPE(RPESC-RPE) Transplantation in Patients with Dry Age-related Macular Degeneration (AMD)

ClinicalTrials.gov ID NCT04627428

Sponsor Luxa Biotechnology, LLC
Information provided by Luxa Biotechnology, LLC (Responsible Party)
Last Update Posted 2023-01-31

Brief Summary
The main objective of the study is evaluation of the safety and tolerability of RPESC-RPE-4 W as therapy for dry AMD.

Detailed Description
RPESC-RPE-4 W is Allogeneic RPE stem cell (RPESC)-derived RPE cells (RPESC-RPE) isolated from the RPE layer of human cadaveric eyes are transplanted under the macular.

This first-in-human Phase 1/2a open-label dose-escalation interventional study plans to enroll a total of 18 subjects.

Official Title
A Phase1/2a, Open-Label Study to Evaluate the Safety and Tolerability of RPE Stem Cell-derived RPE (RPESC-RPE) Transplantation as Therapy for Dry Age-related Macular Degeneration (AMD)

Conditions
Dry Age-related Macular Degeneration

Study Type
Interventional

Enrollment (Estimated)
18

Intervention/Treatment
Biological: RPESC-RPE-4 W

Phase
Phase 1 Phase 2

Other Study ID Numbers
RPESC-RPE-01
U01EY030581 (U.S. NIH Grant/Contract)
UG3EY031810 (U.S. NIH Grant/Contract)

Study Start (Actual)
2022-04-05

Primary Completion (Estimated)
2025-05-31

Study Completion (Estimated)
2025-05-31

Study Contact
Name: Jeffrey H Stern, M.D., Ph.D.
Phone Number: 05184371111
Email: jeffreystern@luxabiotech.com
United States

Ann Arbor, Michigan, United States, 48105
Recruiting
University of Michigan Kellogg Eye Center

Contact
Rajesh C Rao, M.D.

Eligibility Criteria
Ages Eligible for study
55 Years and older (Adult, Older Adult)

Accepts Healthy Volunteer
No

Sexes Eligible for Study
All

Description

Inclusion Criteria

- Clinical diagnosis of dry AMD.
- Ability to understand and give informed consent.
- Adult male or female >55 years of age.
- Medically suitable to undergo vitrectomy and subretinal injection (>60% on Karnofsky scale).
- Postmenopausal if female (expected to be common for the age limitation), or the female partner of a male subject unable to father children.
- If male, willing to use barrier and spermicidal contraception during the study.

Exclusion Criteria

- Allergy or hypersensitivity to dilation drops or fluorescein.
- Active major medical conditions limiting ability to participate in the study.
- Active malignancy or treatment with chemotherapy.
- Systemic immunosuppressant therapy within past six months.
- History of toxoplasmosis, retinal histoplasmosis, or tuberculosis.
- Receipt of investigational product (IP) in a clinical trial within prior six months.
- Any other medical condition, which, in the Investigator's judgment, will interfere with the subject's ability to comply with the protocol, compromises subject safety, or interferes with the interpretation of the study results.
- Pregnant or nursing females.

Design Details

Primary Purpose: Treatment
Allocation: Non-Randomized
Interventional Model: Sequential Assignment
Masking: None (Open Label)

Arms and interventions

Participant group/arm	Intervention/treatment
Experimental: 50,000 cells Six patients will receive single dose of 50,000 RPESC-RPE-4 W cells in the eye	Biological: RPESC-RPE-4 W • RPESC-RPE-4 W
Experimental: 150,000 cells Six patients will receive single dose of 150,000 RPESC-RPE-4 W cells in the eye	Biological: RPESC-RPE-4 W • RPESC-RPE-4 W
Experimental: 250,000 cells Six patients will receive single dose of 250,000 RPESC-RPE-4 W cells in the eye	Biological: RPESC-RPE-4 W • RPESC-RPE-4 W

Primary outcome measures

Outcome measure	Measure description	Time frame
Safety and tolerability of RPESC-RPE-4 W transplantation	The transplantation of RPESC-RPE-4 W cells will be considered safe and tolerated in the absence of: • Decrease in visual acuity (VA) of more than 15 Early Treatment Diabetic Retinopathy Study (ETDRS) letters (or to worse than counting fingers at three feet) from baseline • Any Grade 2 (CTCAE version 5) or greater Adverse Events (AE) related to the cell product and investigational interventions • Any evidence that the cells are contaminated with an infectious agent or serious immune response to the cell product • Any evidence that the cells show tumorigenic potential	24 months

Secondary outcome measures

Outcome measure	Measure description	Time frame
Change in the mean of Best-Corrected Visual Acuity (BCVA)	Change in visual acuity will be measured by ETDRS chart	24 months
Loss of ≥10 decibels of ten-degree average visual sensitivity microperimetry	Loss of ≥10 decibels of ten-degree average visual sensitivity will be measured by microperimetry	24 months
Change in GA lesion area	Change in GA lesion area will be measured	24 months
Evidence of structural changes	Structural evidence will be measured by OCT imaging, autofluorescence, fluorescein angiography, and fundus photography	24 months

Sponsor

Luxa Biotechnology, LLC

Collaborators

- National Institutes of Health (NIH)
- National Eye Institute (NEI)
- Regenerative Research Foundation

Investigators

- Principal Investigator: Rajesh C Rao, M.D., University of Michigan Kellogg Eye Center

General Publications

No publications available

Recruiting

Autologous Transplantation of Induced Pluripotent Stem Cell-Derived Retinal Pigment Epithelium for Geographic Atrophy Associated With Age-Related Macular Degeneration

ClinicalTrials.gov ID NCT04339764

Sponsor National Eye Institute (NEI)
Information provided by National Eye Institute (NEI) (Responsible Party)
Last Update Posted 2023-05-12

Brief Summary

Background
Age-related macular degeneration is a common eye disease in people over 50. The "dry" form of the disease can worsen into geographic atrophy, causing blind spots. Researchers want to learn if replacing older eye cells with younger ones can help treat this disease.

Objective
To test the safety of putting cells inside the eye as a possible future treatment for dry age-related macular degeneration.

Eligibility
People ages 55 and older who have geographic atrophy with loss of vision. People who have had "wet" macular degeneration in either eye are NOT eligible.

Design
Participants will be screened with:

- Medical history
- Physical exam
- Blood and urine tests
- Eye exam
- Eye photos
- Fluorescein angiography. An intravenous (IV) line is placed in an arm vein. A dye is injected. A camera takes pictures of the dye as it flows through the eyes' blood vessels.
- Electroretinography. An electrode is taped to participants' forehead. They sit in the dark. After 30 min, numbing eye drops and contact lenses are placed in their eyes. They watch flashing lights.
- Tuberculosis test
- Chest X-ray
- Electrocardiography. Sticky pads are placed on participants' chest to record the heart's electrical activity.

Participants will have at least 14 study visits over 5 and a half years. They will repeat screening tests.

Participants will have retinal pigment epithelium (RPE) transplantation surgery in one eye. For this, cells from participants' blood are turned into RPE cells. These cells are placed in their eye through a cut in their retina. They will get dilating eye drops, an IV line, and anesthesia that may make them sleep. A gas bubble will be put in their eye to help it heal.

Participants will be contacted yearly for up to 15 years.

Detailed Description

Age-related macular degeneration (AMD) is a leading cause of vision loss among the elderly. There is no treatment for geographic atrophy (GA), the advanced stage of dry AMD, in which cells of the neurosensory retina and associated retinal pigment epithelium (RPE) gradually degenerate and die. Advances in stem cell biology allowing differentiation of pluripotent cells into RPE in vitro make feasible a cell-based strategy for potential treatment of AMD, and recent methods for induced pluripotent stem cell (iPSC) generation offer promise of individualized autologous] therapy. Such an approach involves generation of iPSC from somatic cells taken from a patient with GA, differentiation of iPSC into RPE grown as a monolayer on a thin scaffold in vitro, and transplantation of the RPE/scaffold construct into a small region in the subretinal space of the same patient, with a goal of rescuing the overlying neurosensory retina from further degeneration.

Objective

To evaluate the safety and feasibility of subretinal transplantation of iPSC-derived RPE, grown as a monolayer on a biodegradable poly lactic-co-glycolic acid (PLGA) scaffold, as a potential autologous cell-based therapy for GA associated with AMD.

Study Population: Five participants will undergo RPE transplantation in one eye. Eligible eyes will have GA, best-corrected visual acuity (BCVA) between 20/100 and inclusive of counting fingers (CF), and a fellow eye that has same or better BCVA. If the National Eye Institute (NEI) Data and Safety Monitoring Committee (DSMC) gives clearance to proceed based on review of data from the first cohort, a second cohort of up to seven additional participants with GA, BCVA between 20/80, and CF (inclusive) in the eye being considered for RPE transplantation, and same or better visual acuity in the other eye may undergo the procedure to gather additional safety and potential efficacy data useful for planning future studies. Up to 20 participants may be enrolled to allow for screening failures, for participants withdrawing from the study prior to RPE transplantation, or cases where the RPE cell transplantation does not occur due to intraoperative surgical considerations.

Design

In this phase I/IIa, prospective, single-arm, single-center clinical trial, participants will undergo subretinal transplantation of autologous iPSC-derived RPE in one eye and will be followed for five years after surgery.

Outcome Measures

The primary outcome measure is the safety of RPE/PLGA transplantation, as determined by assessment of visual acuity change and summary of adverse events at 12 months after RPE/PLGA transplantation. Secondary outcome measures include visual acuity change and adverse event reporting at 24 and 60 months, and changes in the following at 12, 24, and 60 months as compared with baseline, assessed in the transplanted region, and compared where applicable with other areas in the macula, and/or with corresponding regions in the fellow eye: retinal sensitivity and fixation parameters assessed by microperimetry; multifocal electroretinography (mfERG) responses; macular structure on cross-sectional and en face imaging by optical coherence tomography (OCT); macular features on color, single-wavelength reflectance, and fundus autofluorescence (FAF) photography; and fluorescein angiography (FA). Some NEI participants may undergo imaging of photoreceptor/RPE features using adaptive-optics-assisted macular imaging under a separate protocol (e.g., 15-EI-0020).

Official Title
A Phase I/IIa Trial for Autologous Transplantation of Induced Pluripotent Stem Cell-Derived Retinal Pigment Epithelium for Geographic Atrophy Associated With Age-Related Macular Degeneration

Conditions
Age-Related Macular Degeneration

Study Type
Interventional

Enrollment (Estimated)
20

Intervention/Treatment
Drug: iPSC-derived RPE/PLGA transplantation

Phase
Phase 1 Phase 2

Other Study ID Numbers
200052
20-EI-0052

Study Start (Actual)
2020-09-23

Primary Completion (Estimated)
2029-05-31

Study Completion (Estimated)
2029-05-31

Study Contact
Name: Angel H Garced, R.N.
Phone Number: (301) 594-3141
Email: angel.garced@nih.gov

Study Contact Backup
Name: M. Teresa Magone de Quadros Costa, M.D.
Phone Number: (301) 435-4562
Email: teresa.magonedequadroscosta@nih.gov

United States

Bethesda, Maryland, United States, 20892
Recruiting
National Institutes of Health Clinical Center

Contact
For more information at the NIH Clinical Center contact Office of Patient Recruitment (OPR)
800-411-1222 ext. TTY8664111010prpl@cc.nih.gov

Eligibility Criteria

Ages Eligible for Study
55 Years and older (Adult, Older Adult)

Accepts Healthy Volunteer
No

Sexes Eligible for Study
All

Description

Inclusion Criteria
- To be eligible, the following inclusion criteria must be met, where applicable.
- Participant must be 55 years of age or older.
- Participant must have a diagnosis of AMD, defined as presence (or history, as documented in available color fundus photographs) of at least one medium or large druse (greater than or equal to 63 micrometer diameter) in the macula in at least one eye; AND presence of GA in at least one eye.
- Participant must understand and sign the protocol's informed consent document.
- Any participant of childbearing potential must have a negative pregnancy test at screening and must be willing to undergo pregnancy testing prior to RPE transplantation.
- Any participant of childbearing potential and any participant able to father children must have (or have a partner who has) had a hysterectomy or vasectomy, be completely abstinent from intercourse, or must agree to practice an effective method of contraception through Month 12 in the study. Acceptable methods of contraception include:
  - Hormonal contraception (i.e., birth control pills, injected hormones, dermal patch or vaginal ring),
  - Intrauterine device,

  - Barrier methods (diaphragm, condom) with spermicide, or
  - Surgical sterilization (tubal ligation).

- Participant must be medically able to comply with the study treatment (including ability to safely receive anesthesia for surgery), study testing and procedures, and follow-up visits.

Study Eye Inclusion Criteria

- The study eye must have one or more regions of geographic atrophy with total area of 1 disc area or more. A region of geographic atrophy is defined as an area of uniform hypofluorescence on fundus autofluorescence (FAF) imaging, with greatest linear dimension at least 500 micrometer, with a border within 500 micrometer of the foveal center, not compatible with pigmentary changes, drusen, RPE detachment, drusenoid RPE detachment, hemorrhage, or other lesion. (Note: If macular geographic atrophy is contiguous with peripapillary atrophy, complicating calculation of total area, only atrophy temporal to a vertical line placed a half disc diameter temporal to the temporal border of the disc will be included in the total area of geographic atrophy calculated for eligibility purposes.)
- For participants in the first cohort, the study eye must have an ETDRS best-corrected visual acuity (BCVA) letter score of less than or equal to 53 and greater than or equal to CF (i.e., Snellen equivalent between 20/100 and CF), and the fellow eye must have a letter score no more than five letters worse than the study eye using Electronic Visual Acuity (EVA) testing. (Note: Letter scores within five or fewer letters of each other are accordingly considered equal for eligibility determination, and other factors may be used to select the study eye if both are eligible by BCVA.)
- For participants in the second cohort, the study eye must have an ETDRS best-corrected visual acuity (BCVA) letter score of less than or equal to 58 and greater than or equal to CF (i.e., Snellen equivalent between 20/80 and CF), and the fellow eye must have a letter score no more than five letters worse than the study eye using Electronic Visual Acuity (EVA) testing. (Note: Letter scores within five or fewer letters of each other are accordingly considered equal for eligibility determination, and other factors may be used to select the study eye if both are eligible by BCVA.)
- The compromise in visual acuity for the study eye must be judged predominantly secondary to dry AMD, in the judgment of the investigator.
- The study eye must have clarity of ocular media and degree of pupil dilation sufficient to permit adequate fundus photography and safe vitrectomy surgery.
- The study eye must be either pseudophakic or aphakic.

Exclusion Criteria

A participant is not eligible if any of the following exclusion criteria are present:

- Participant is actively receiving another study medication/investigational product (IP).
- Participant has any condition that significantly increases risk of systemic corticosteroids or systemic steroid-sparing immuno-modulatory agents, such as uncon-

trolled diabetes mellitus, chronic hepatitis or liver failure, chronic renal failure, or present infection with HIV, syphilis, tuberculosis, hepatitis B, or hepatitis C (past infection now resolved, where applicable, is not exclusionary; but persistent infection, even if latent, is exclusionary).
- Participant has diagnosis of a malignancy expected to affect two-year survival.
- Participant is pregnant, breast-feeding, or planning to become pregnant through the first 12 months of the study.
- Participant has a family history of a retinal degeneration other than AMD suspected to play a role in the ocular phenotype of the participant in the judgment of the investigator, based on disease features and mode of inheritance, such as in a case of autosomal dominant retinal degeneration in a parent or child.
- Participant is taking, or has taken within the previous year, medication with known potential toxicity to the retina, optic nerve, or lens (such as chloroquine, hydroxychloroquine, ethambutol).
- Participant is unable or unwilling to give informed consent that includes use of medical records and clinical samples for current and future research.

Study/Eye Exclusion Criteria

- The study eye must not have macular subretinal or choroidal neovascularization, as assessed by FA and OCT; or any history of such neovascularization (as assessed by past available records or images).
- The study eye must not have any serous or hemorrhagic pigment epithelial detachment, as assessed by FA and OCT.
- The study eye must not have any history of photodynamic therapy (PDT) or macular thermal laser photocoagulation, or history of intravitreal injection of anti-vascular endothelial growth factor (VEGF) agents or corticosteroids (excepting medications used perioperatively at prior cataract surgery).
- The study eye must not have an axial length > 25.0 mm.
- The study eye must not have had any surgery in the previous 12 weeks or laser capsulotomy in the previous four weeks.
- The study eye must not have chronic glaucoma; OR significant ocular hypertension, defined as documented intraocular pressure of greater than or equal to 26 mmHg on at least two occasions in the absence of self-limited acute glaucoma; OR history of probable or definite steroid response manifesting as acute glaucoma or ocular hypertension, even if self-limited and no longer present; and the fellow eye must not have evidence for present or past glaucoma or ocular hypertension judged to significantly impact the risk of glaucoma in the study eye (including history of probable or definite steroid response). (Note: History of self-limited acute glaucoma in a study or fellow eye, if not secondary to steroid response, and if now resolved and not expected to recur (e.g., history of elevated intraocular pressure from retained viscoelastic after cataract surgery), is not exclusionary. History of glaucoma or ocular hypertension in the fellow eye, if not felt to significantly impact risk of glaucoma in the study eye, is not exclusionary.)
- The study eye must not have a condition materially increasing the risks of surgery or potentially affecting visual function over the next two years in the judg-

ment of the investigator, such as chronic uveitis, diabetic retinopathy, keratitis, scleritis, optic neuropathy, untreated retinal detachment, macular edema from prior vein occlusion or other cause, proliferative vitreoretinopathy (PVR), vitreous hemorrhage, pathologic myopia, etc. A history of such conditions is not exclusionary, if judged to not materially increase risks of surgery or to potentially affect vision in the next 2 years in the opinion of the investigator.

Design Details
Primary Purpose: Treatment
Allocation: N/A
Interventional Model: Single Group Assignment
Masking: None (Open Label)

Arms and interventions

Participant group/arm	Intervention/treatment
Experimental: Participants receiving intervention Participants receiving intervention	Drug: iPSC-derived RPE/PLGA transplantation • iPSC-derived RPE/PLGA transplantation

Primary outcome measures

Outcome measure	Measure description	Time frame
Visual acuity change	Safety measure	12, 24, and 60 months
Summary of adverse events	Safety measure	12, 24, and 60 months

Secondary outcome measures

Outcome measure	Measure description	Time frame
Retinal Structure (optical coherence tomography)	Safety and efficacy measure	12, 24, and 60 months
Retinal sensitivity and fixation (microperimetry)	Safety and efficacy measure	12, 24, and 60 months
Multifocal electroretinography responses	Safety and efficacy measure	12, 24, and 60 months
Retinal Structure (color and autofluorescence imaging)	Safety and efficacy measure	12, 24, and 60 months
Retinal structure (fluorescein angiography)	Safety and efficacy measure.	12, 24, and 60 months

Sponsor
National Eye Institute (NEI)

Collaborators
No information provided

Investigators
- Principal Investigator: M. Teresa Magone de Quadros Costa, M.D., National Eye Institute (NEI)

General Publications
No publications available

Enrolling By Invitation

Clinical Trial of Autologous Intravitreal Bone-Marrow CD34+ Stem Cells for Retinopathy

ClinicalTrials.gov ID NCT01736059

Sponsor University of California, Davis
Information provided by University of California, Davis (Responsible Party)
Last Update Posted 2022-08-09

Brief Summary
This pilot study is to determine whether it would be safe and feasible to inject CD34+ stem cells from bone marrow into the eye as treatment for patients who are irreversibly blind from various retinal conditions.

Detailed Description
In this pilot clinical trial, eyes with irreversible vision loss from retinal degenerative conditions (macular degeneration or retinitis pigmentosa) or retinal vascular disease (diabetic retinopathy or retinal vein occlusion) will be treated with intravitreal injection of autologous CD34+ stem cells isolated from bone marrow aspirate under Good Manufacturing Practice conditions. This study will determine whether there are any major safety and feasibility concerns using this therapy. Patients will be followed for 6 months after treatment by serial comprehensive eye examination supplemented with various retinal imaging and diagnostic tests.

Official Title
A Pilot Clinical Trial of the Feasibility and Safety of Intravitreal Autologous Adult Bone Marrow Stem Cells in Treating Eyes With Vision Loss From Retinopathy

Conditions
Non-exudative Age-related Macular Degeneration
Diabetic Retinopathy
Retina Vein Occlusion
Retinitis Pigmentosa
Hereditary Macular Degeneration

Study Type
Interventional

Enrollment (Estimated)
15

Intervention/Treatment
Drug: CD34+ bone marrow stem cells intravitreal

Phase
Phase 1

Other Study ID Numbers
906595

Study Start
2012-07

Primary Completion (Estimated)
2023-06

Study Completion (Estimated)
2023-06
United States

Sacramento, California, United States, 95817
University of California Davis

Eligibility Criteria
Ages Eligible for Study
18–100 Years (Adult, Older Adult)

Accepts healthy volunteer
No

Sexes Eligible for Study
All

Description

Inclusion Criteria
Age > 18 years of age
Visual acuity 20/100 to CF.
Duration of vision loss >3 months
Vision loss from macular degeneration, retinitis pigmentosa, retinal vein occlusion, or diabetic retinopathy
The study eye has the worse visual acuity
No active eye or systemic disease
No history of macular edema or retinal/choroidal neovascularization requiring treatment within 6 months
No significant media opacity
No coagulopathy or other hematologic abnormality
No concurrent immunosuppressive therapy
Able to keep follow-up for 6 months

Exclusion Criteria

Allergy to fluorescein dye
Other concurrent retinal or optic nerve disease affecting vision

Design Details

Primary Purpose: Treatment
Allocation: N/A
Interventional Model: Single Group Assignment
Masking: None (Open Label).

Arms and interventions

Participant group/arm	Intervention/treatment
Experimental: Stem cell treated	Drug: CD34+ bone marrow stem cells intravitreal

Primary outcome measures

Outcome measure	Measure description	Time frame
Incidence and severity of ocular adverse events	As evident on eye examination by severe decrease in vision and/or adverse effect requiring major treatment intervention directly attributable to study treatment	1 day to 6 months

Secondary outcome measures

Outcome measure	Measure description	Time frame
The number of stem cells isolated and injected into the study eye	Will determine the yield of the bone marrow aspiration and stem cell isolation procedures	1 day

Other outcome measures

Outcome measure	Measure description	Time frame
Incidence and severity of systemic adverse events	Any systemic adverse event attributable to the bone marrow aspiration or intravitreal injection of the stem cells	1 day to 6 months

Sponsor

University of California, Davis

Collaborators

No information provided

Investigators

- Principal Investigator: Susanna S Park, MD PhD, University of California Davis Eye Center.

General Publications

- Park SS. Cell Therapy Applications for Retinal Vascular Diseases: Diabetic Retinopathy and Retinal Vein Occlusion. Invest Ophthalmol Vis Sci. 2016 Apr 1;57(5):ORSFj1–ORSFj. https://doi.org/10.1167/iovs.15-17594.

Chapter 2
Active, Not Recruiting Ocular Stem Cell Studies

Safety and Tolerability of hRPC in Retinitis Pigmentosa (hRPCRP)

ClinicalTrials.gov ID NCT02464436

Sponsor ReNeuron Limited
Information provided by ReNeuron Limited (Responsible Party)
Last Update Posted 2023-07-06

Brief Summary
hRPC is a cell therapy for retinitis pigmentosa. This is a first-in-human, dose-escalation study in which participants with retinitis pigmentosa will receive a single subretinal injection of hRPC cells in one eye to evaluate safety and tolerability.

Participants will be followed for 2 years to evaluate the safety and tolerability of hRPC. Additional testing will seek to establish any preliminary efficacy from hRPC.

Detailed Description
This is a first-in-human open-label phase I/II dose-escalation study in which participants with retinitis pigmentosa will receive a single uniocular subretinal implantation of one of three doses of hRPC.

Treated eyes will be carefully monitored for any ocular or systemic adverse events for 2 years.

Testing will comprise a series of detailed ophthalmic examinations and imaging together with blood testing and systemic evaluations, as necessary.

Ophthalmic testing will also be evaluated for any preliminary efficacy signal.

Official Title
First-in-human Phase I/IIa, Open-Label, Prospective Study of the Safety and Tolerability of Subretinally Transplanted Human Retinal Progenitor Cells (hRPC) in Patients With Retinitis Pigmentosa (RP)

J. N. Weiss, *Ocular Stem Cell Studies*,
https://doi.org/10.1007/978-3-031-58557-9_2

Conditions
Retinitis Pigmentosa

Intervention/Treatment
Drug: hRPC
United States
Arizona Locations

Phoenix, Arizona, United States, 85053
Retinal Research Institute

Massachusetts Locations

Boston, Massachusetts, United States, 02114
Massachusetts Eye and Ear Infirmary

Oregon Locations

Portland, Oregon, United States, 97239
Oregon Health and Science University

Spain

Barcelona, Spain
Institut de la Màcula

United Kingdom

Oxford, United Kingdom, OX3 9DU
Oxford Eye Hospital

Eligibility Criteria
Ages Eligible for Study
18 years and older (adult, older adult)

Accepts Healthy Volunteer
No

Sexes Eligible for Study
All

Description

Inclusion Criteria
- Have ability to give written informed consent as evidenced by signature on the subject consent form.
- Be adult male or female over 18 years of age.
- Have clinical diagnosis of RP, based upon one or more of the following: clinical features, medical imaging, electrophysiological measures, and genetic testing, if available. Genetic confirmation is not obligatory.
- Have best-corrected ETDRS visual acuity of 35 letters or less (approximately 20/200 or worse) in the study eye for cohorts 1–5; have best-corrected ETDRS

visual acuity of 63 letters (approximately 20/63) to 36 letters (approximately 20/200) in the study eye for cohorts 6–8, and best-corrected ETDRS visual acuity of 8 letters (approximately 20/800) to 68 letters (approximately 20/50) for cohorts 9 and so on.

- Be able to complete the entire microperimetry test and demonstrate adequate fixation and consistency between baseline readings such that the accuracy of both baseline and follow on testing should enable the detection of clinically significant changes in retinal sensitivity.
- Be medically able to undergo vitrectomy and subretinal injection.
- Have good general health as defined by:
  - Normal serum chemistry and hematology. Out of normal range laboratory findings deemed not clinically significant are acceptable.
  - No history of malignancy, except non-melanoma skin cancer; pre-malignant conditions, and cancer in situ.
  - Negative serology for human immunodeficiency virus (HIV), hepatitis B (HBV), hepatitis C (HCV).
  - Medically fit enough to undertake surgery which may require general anesthesia as well as medically fit to undergo a short perioperative course of systemic corticosteroid therapy.
  - Free of any other systemic condition that in the opinion of the investigator may have an impact on the safety of the subject, conduct of study procedures, or integrity of study data (e.g., severe cardiovascular or respiratory disease; poorly controlled diabetes; significant psychiatric impairment).
- Females of childbearing potential must have a confirmed negative pregnancy test at visits 1 and 3 and be willing to use highly effective method of contraception (e.g., oral contraceptive and condom, intrauterine device (IUD) and condom, diaphragm with spermicide and condom) for the duration of this study.
- Males must be willing to use a reliable method of contraception (e.g., barrier and spermicide) for the duration of this study; unless have been surgically sterilized with confirmed azoospermia.
- Be willing and able to attend all scheduled clinical assessments, ability to communicate well with the investigator, and to comply with the expectations of the study.

Exclusion Criteria

- Exhibits a difference in ETDRS BCVA of 15 letters of more in either eye between any of the baseline visits.
- Exhibits a difference in ETDRS BCVA of 20 or more letters between eyes at the time of any of the screening or baseline visits attributed to asymmetry in the progression of RP.
- Presence of ocular disease or ocular media opacity in the study eye, which in the opinion of the investigator, will preclude an accurate evaluation at any time during the study.
- History of any retinal and/or macular disease other than RP (e.g., retinal detachment) that in the opinion of the investigator may have an impact on the safety of

the subject, conduct of study procedures, or integrity of study data. Specifically, subjects in whom significant preexisting vitreoretinal pathology might influence visual acuity outcomes should be excluded.
- Active ocular infection or inflammation, or any history of intraocular inflammation, that would expose subject to risk during or following surgery.
- Prior vitrectomy in the study eye.
- A history of amblyopia in the study eye.
- High myopia (>6 diopters) in the study eye.
- Cataract surgery in the study eye or ocular surgery in either eye (which in the opinion of the investigator may have an impact on patient safety or the integrity of data from the study eye) during the study or within 3 months prior to treatment.
- Participation in any clinical study involving an investigational drug or device within 6 months prior to treatment or 5 half-lives of the drug (whichever is longer) prior to initiation of treatment.
- Prior stem cell administration or injections to any part of the body (subjects who have received autologous bone marrow stem cell transplant will be eligible).
- Use of systemic immunosuppressive agents (e.g., corticosteroid) in the 6 months prior to treatment or 5 half-lives of the drug (whichever is longer) prior to initiation of treatment (note: inhaled, intranasal, and/or topical dermatologic steroids are allowed).
- (For females) Be breastfeeding or planning a pregnancy.

Design Details
Primary Purpose: Treatment
Allocation: N/A
Interventional Model: Single-Group Assignment
Masking: None (open label)

Arms and interventions

Participant group/arm	Intervention/treatment
Experimental: human retinal progenitor cells (hRPC) Single subretinal administration of human retinal progenitor cells (hRPC)	Drug: hRPC • Participants will undergo vitrectomy surgery and subretinal implantation of hRPC in the study eye • Other Names: – Human retinal progenitor cells

Primary outcome measures

Outcome measure	Measure description	Time frame
Safety over the 6 months after treatment as assessed by the incidence of treatment-emergent adverse events (TEAEs) and changes from baseline in other safety parameters	Safety measures will be assessed by review of important events, including but not limited to inflammation, complications of the surgical procedure and worsening of vision	6 months

Secondary outcome measures

Outcome measure	Measure description	Time frame
Safety (visual function measure: change in visual acuity)	A summary of the ETDRS ± BRVT BCVA letter score and the change from baseline to end of study in the treated eye presented by treatment group	24 months
Safety (visual function measure: change in visual field: Goldmann visual field, microperimetry, and FST)	A summary of the perimetry and change from baseline to end of study in the treated eye presented by treatment group	24 months
Safety (change in retinal sensitivity in the area overlying the implanted hRPC as compared with untreated retina)	ERG results and change from baseline to end of study summarized descriptively and presented by treatment group	24 months
Safety (anatomical endpoint relating to retinal function in implant location—color fundus photography)	A qualitative description of the change in retinal appearance of treated and untreated retinal area in the treated eye from baseline to end of study presented by treatment group	24 months
Safety (anatomical endpoint relating to retinal function in implant location—fundus autofluorescence)	A qualitative description of the change in retinal appearance of treated and untreated retinal area in the treated eye from baseline to end of study presented by treatment group	24 months
Safety (anatomical endpoint relating to retinal function in implant location—spectral domain-OCT)	Summary of the overall retinal thickness, outer nuclear layer thickness, and ellipsoid zone measurement and change from baseline to the end of study of treated retina compared with untreated retina in the treated eye by treatment group	24 months

Sponsor

ReNeuron Limited

Collaborators

No information provided

Investigators

- Principal Investigator: Jason Comander, MD, Massachusetts Eye and Ear Infirmary (MEEI)
- Study Director: Vince Holmes, ReNeuron Limited

General Publications

No publications available.

Belarus

Enrolling by Invitation

Treatment of Age-Related Macular Degeneration Using Retinal Stem and Progenitor Cells

ClinicalTrials.gov ID NCT05187104

Sponsor Institute of Biophysics and Cell Engineering of National Academy of Sciences of Belarus
Information provided by Institute of Biophysics and Cell Engineering of National Academy of Sciences of Belarus (Responsible Party)
Last Update Posted 2022-09-14

Brief Summary
Treatment of age-related macular degeneration using retinal stem and progenitor cells.

Detailed Description
The aim of the project is to develop a biomedical cell product based on retinal pigment epithelium cells for highly effective treatment age-related macular degeneration; to conduct the clinical trials of the biomedical cell product in the treatment of age-related macular degeneration.

Official Title
Treatment of age-related macular degeneration using retinal stem and progenitor cells

Conditions
Age-related macular degeneration

Study Type
Interventional

Enrollment (Estimated)
20

Intervention/Treatment
Biological: Cultured retinal stem and progenitor cells
Other: Standard treatment according to the clinical protocols

Phase
Phase 1 Phase 2

Study Start (Actual)
2022-03-01

Primary Completion (Estimated)
2023-12-31

Study Completion (Estimated)
2023-12-31

Belarus
Minsk, Belarus, 220072
Institute of Biophysics and Cell Engineering of National Academy of Sciences of Belarus

Eligibility Criteria
Ages Eligible for Study
18–80 years (adult, older adult)

Accepts Healthy Volunteer
No

Sexes Eligible for Study
All

Description

Inclusion Criteria
- Clinical diagnosis of age-related macular degeneration
- The patient can read, understand, follow the examination procedures, and complete, if necessary, the required documentation
- Written informed consent

Exclusion Criteria
- The presence of any malignant tumor within the last 5 years
- Acute or chronic diseases in the stage of decompensation
- Chronic infectious diseases: HIV, viral hepatitis B, C, tuberculosis
- Patients who are pregnant, breastfeeding, or fertile patients who are not using adequate contraceptive methods
- Chronic and protracted mental disorders, all diseases with the presence of the syndrome of dependence on alcohol, drugs, and psychoactive substances, any other condition that makes the patient unable to understand the nature, extent, and possible consequences of the study or, in the opinion of the researcher, prevents the patient from observing and performing protocol
- Patients are unable or unwilling to give written informed consent and/or follow research procedures
- Any other medical condition that, in the opinion of the investigator, may be associated with an increased risk to the patient or may affect the outcome or evaluation of the study

Design Details
Primary purpose: treatment
Allocation: non-randomized
Interventional model: parallel assignment
Masking: None (open label)

Arms and interventions

Participant group/arm	Intervention/treatment
Experimental: Patients with age-related macular degeneration receiving standard treatment and retinal stem cells Patients with age-related macular degeneration receiving standard treatment and autologous retinal stem and progenitor cells	Biological: Cultured retinal stem and progenitor cells • Cultured retinal stem and progenitor cells injected subretinally Other: Standard treatment according to the clinical protocols • Standard treatment of age-related macular degeneration according to the clinical protocols
Active comparator: Patients with age-related macular degeneration receiving standard treatment Patients with age-related macular degeneration receiving standard treatment	Other: Standard treatment according to the clinical protocols • Standard treatment of age-related macular degeneration according to the clinical protocols

Primary outcome measures

Outcome measure	Measure description	Time frame
Visual acuity	Determination of visual acuity after the therapy	1 month
Visual acuity	Determination of visual acuity after the therapy	1 year
Adverse effects associated with the therapy	Determination of adverse effects associated with the therapy	1 month
Adverse effects associated with the therapy	Determination of adverse effects associated with the therapy	1 year

Sponsor

Institute of Biophysics and Cell Engineering of National Academy of Sciences of Belarus

Collaborators

- Belarusian State Medical University

Investigators

- Study Director: Ludmila Marchenko, Prof, Belarusian State Medical University
- Study Director: Zinaida Kvacheva, Dr, The Institute of Biophysics and Cell Engineering of National Academy of Sciences of Belarus

General Publications

No publications available.

Caribbean/Argentina/Mexico

Recruiting

Safety of Cultured Allogeneic Adult Umbilical Cord-Derived Mesenchymal Stem Cells for Eye Diseases

ClinicalTrials.gov ID NCT05147701

Sponsor The Foundation for Orthopaedics and Regenerative Medicine
Information provided by The Foundation for Orthopaedics and Regenerative Medicine (Responsible Party)
Last Update Posted 2022-10-27

Brief Summary
This trial will study the safety and efficacy of intravenous and subtenon delivery of cultured allogeneic adult umbilical cord-derived mesenchymal stem cells for the treatment of Eye diseases.

Detailed Description
This patient-funded trial aims to study the safety and efficacy of intravenous and subtenon delivery of cultured allogeneic adult umbilical cord-derived mesenchymal stem cells (UC-MSCs) for the treatment of eye diseases. The total dose will be 100 million cells. Patients will be evaluated within 1 month pre-treatment and at 1, 6, 12, 24, 36, and 48 months post-treatment for safety and efficacy.

For patients with more severe disease, an autologous effector cells (activated lymphocytes) treatment will be utilized created from the patient's own cells obtained by apheresis.

Official Title
Safety of Cultured Allogeneic Adult Umbilical Cord-Derived Mesenchymal Stem Cells for the Treatment of Eye Diseases

Conditions
Eye Diseases
Retinitis Pigmentosa
Glaucoma
Diabetic Retinopathy
Macular Degeneration
Traumatic Optic Neuropathy
Optic Atrophy

Study Type
Interventional

Enrollment (Estimated)
20

Intervention/Treatment
Biological: AlloRx

Phase
Phase 1

Other Study ID Numbers
ATG-1-MSC-014

Study Start (Actual)
2022-02-01

Primary Completion (Estimated)
2026-01

Study Completion (Estimated)
Antigua and Barbuda

St. John's, Antigua and Barbuda
Recruiting
Medical Surgical Associates Center

Contact
Chadwick Prodromos, M.D.
8476996810 care@thepsci.com

Argentina

Buenos Aires, Argentina
Recruiting
Center for Investigation in Tissue Engineering and Cellular Therapy

Contact
Chadwick Prodromos, M.D.
8476996810 ext 202 Care@thepsci.com

Mexico
N.L Locations

San Pedro Garza García, N.L, Mexico
Recruiting
Medyca Bosques

Contact
Chadwick Prodromos, M.D.
8476996810 ext 202 Care@thepsci.com

Eligibility Criteria
Ages Eligible for Study
(Child, adult, older adult)

Accepts Healthy Volunteer
No

Sexes Eligible for Study
All

Description

Inclusion Criteria
- Diagnosis of eye disease
- Understanding and willingness to sign a written informed consent document

Exclusion Criteria
- Active infection
- Active cancer
- Chronic multisystem organ failure
- Pregnancy
- Clinically significant abnormalities on pre-treatment laboratory evaluation
- Medical condition that would (based on the opinion of the investigator) compromise patient's safety.
- Continued drug abuse
- Pre-menopausal women not using contraception
- Previous organ transplant
- Hypersensitivity to sulfur

Design Details
Primary purpose: treatment
Allocation: N/A
Interventional model: single-group assignment
Masking: none (open label)

Arms and interventions

Participant group/arm	Intervention/treatment
Experimental: Treatment Group (AlloRx) intravenous and subtenon delivery (total dose of 100 million cells)	Biological: AlloRx • Cultured allogeneic adult umbilical cord-derived mesenchymal stem cells

Primary outcome measures

Outcome measure	Measure description	Time frame
Safety (adverse events)	Clinical monitoring of possible adverse events or complications	Four-year follow-up

Sponsor
The Foundation for Orthopaedics and Regenerative Medicine

Collaborators
No information provided

Investigators

- Principal Investigator: Chadwick Prodromos, MD, The Foundation for Orthopaedics and Regenerative Medicine

General Publications

- Sung Y, Lee SM, Park M, Choi HJ, Kang S, Choi BI, Lew H. Treatment of traumatic optic neuropathy using human placenta-derived mesenchymal stem cells in Asian patients. Regen Med. 2020;15(10):2163–2179. https://doi.org/10.2217/rme-2020-0044. Epub 2020 Dec 14.
- Zhao T, Liang Q, Meng X, Duan P, Wang F, Li S, Liu Y, Yin ZQ. Intravenous Infusion of Umbilical Cord Mesenchymal Stem Cells Maintains and Partially Improves Visual Function in Patients with Advanced Retinitis Pigmentosa. Stem Cells Dev. 2020;29(16):1029–1037. https://doi.org/10.1089/scd.2020.0037. Epub 2020 Jul 15.
- Ozmert E, Arslan U. Management of retinitis pigmentosa by Wharton's jelly derived mesenchymal stem cells: preliminary clinical results. Stem Cell Res Ther. 2020;11(1):25. https://doi.org/10.1186/s13287-020-1549-6.

China

Recruiting

Safety and Efficacy of Autologous Transplantation of iPSC-RPE in the Treatment of Macular Degeneration

ClinicalTrials.gov ID NCT05445063

Sponsor Beijing Tongren Hospital
Information provided by Beijing Tongren Hospital (Responsible Party)
Last Update Posted 2022-07-06

Brief Summary

This project intends to perform autologous transplantation of induced pluripotent stem cell-derived retinal pigment epithelium (iPSC-RPE). The clinical-grade RPE will be transplanted into subretinal space to treat refractory age-related macular degeneration. The efficacy and safety of RPE transplants to treat macular degeneration will be monitored and analyzed with results from EDTRS, BCVA, OCT, ERG, microperimetry, and fluorescein angiography, before and after the treatment.

Detailed Description

The investigators will generate iPSC lines from recruited participants and differentiate the iPSC into RPE. Autologous iPSC-derived RPE will be transplanted into participants' eyes by subretinal injections. The safety and efficacy will be closely monitored and analyzed.

Official Title
Safety and Efficacy of Autologous Transplantation of Induced Pluripotent Stem Cell-Derived Retinal Pigment Epithelium in the Treatment of Macular Degeneration

Conditions
Macular Degeneration

Study Type
Interventional

Enrollment (Estimated)
10

Intervention/Treatment
Biological: Autologous iPSC-derived RPE

Phase
Phase 1

Other Study ID Numbers
2021-2-14

Study Start (Estimated)
2022-08

Primary Completion (Estimated)
2026-07

Study Completion (Estimated)
2026-12

Study Contact
Name: Xiaohui Zhang
Phone Number: +0086-(010)58265915
Email: zhangxh711@126.com
China

Beijing, Beijing, China, 100730
Recruiting
Beijing Tongren Hospital, Capital Medical University

Contact
Zi-Bing Jin, Doctor
+0086-(010)58265913 jinzibing@foxmail.com

Eligibility Criteria
Ages Eligible for Study
50–75 years (adult, older adult)

Accepts Healthy Volunteer
No

Sexes Eligible for Study
All

Description

Inclusion Criteria
- Aged 50–75 years.
- Clinical diagnosis is consistent with the definition of late dry AMD in the age-related eye disease study (AREDS), with one or more >250 um geographic atrophy in the fovea.
- Clinical diagnosis is wet AMD, but no obvious efficacy after conventional treatment.
- The BCVA of the target eye will be 0.05 to 0.3.
- Voluntary as test subjects, informed consent, regular follow-up on time.

Exclusion Criteria
- One-eyed subjects.
- Macular atrophy caused by other diseases in addition to AMD.
- Suffer from retinitis pigmentosa, choroidal retinitis, central serous choroiditis, diabetic retinopathy, or other retinal vascular and degenerative diseases besides AMD.
- Lens opacities (affecting the central vision), glaucoma, uveitis, retinal detachment, optic neuropathy, and other ocular histories.
- Other intraocular surgery histories besides cataract surgery.
- Combined with severe systemic diseases, such as heart failure, liver disease, renal insufficiency, cor pulmonale, COPD in the previous 12 months.
- Combined with severe infectious diseases, such as HIV, HBV, HCV, syphilis, and tuberculosis.
- Abnormal blood coagulation function or other laboratory tests.
- If female and of childbearing potential, pregnant, breastfeeding, or planning to become pregnant through the study.
- If male, refuse to use barrier and spermicide contraception during the study.
- Malignant tumor and history of malignancy.
- Any immune deficiency.
- Allergy to tacrolimus or other macrolides.
- Any immune deficiency.
- Use glucocorticoids, immunosuppressive drugs, or antipsychotic drugs in the previous 3 months.
- Use anticoagulant or the platelet function is still not restored to normal after stopping antiplatelet drugs for 10 days.
- A history of addiction to alcoholism or prohibited drugs.
- Be participating in other intervention clinical trials or receiving other study medications.
- Poor compliance, difficulty to complete the study, or refusal to informed consent.
- Some other situations which might increase the risks of the subjects or interfere with clinical trials, such as mental disorders, cognitive dysfunction, etc.

Design Details

Primary purpose: treatment
Allocation: N/A
Interventional model: single-group assignment
Masking: none (open label)

Arms and interventions

Participant group/arm	Intervention/treatment
Experimental: Participants receiving intervention Participants will receive autologous transplantation of induced pluripotent stem cell-derived retinal pigment epitheliums	Biological: autologous iPSC-derived RPE • Autologous transplantation of iPSC-derived RPE

Primary outcome measures

Outcome measure	Measure description	Time frame
Safety measure	Safety will be assessed by Adverse Events (AEs) of special interest in regards to the investigational product. This will include obtaining information about Serious Adverse Events (SAEs) that are neurologic, infectious, hematologic or fatal, any AE that causes the subject to withdraw from the study, any new diagnosis of an ocular or immune-mediated disorder, cancer (irrespective of prior history), ectopic or proliferative cell growth (RPE or non-RPE) with adverse clinical consequence, unexpected, clinically significant AE possibly related to the cell transplant procedure or the investigational product (autologous iPSC-RPE), pregnancy in a female subject or the partner of a male subject and pregnancy outcome	12 months

Secondary outcome measures

Outcome measure	Measure description	Time frame
Best-Corrected Visual Acuity (BCVA)	Change in visual acuity will be measured by Early Treatment Diabetic Retinopathy Study (ETDRS) chart	12 months
Optical coherence tomography (OCT) imaging	The number of patients with serious retinal detachment, retinal hemorrhage, and cystoid macular edema, and the change in target treatment areas	12 months
Color and autofluorescence imaging	Change in target treatment areas	12 months
Fluorescein angiography	Change in target treatment areas	12 months
Fundus autofluorescence	Change in target treatment areas	12 months
Microperimetry	Exploratory evaluations for the change of retinal sensitivity from baseline in the region of interest	12 months
Electroretinography (ERG)	Exploratory evaluations for the change of retinal electrophysiology responses from baseline	12 months

Sponsor
Beijing Tongren Hospital

Collaborators
No information provided

Investigators
No information provided

General Publications
No publications available.

Great Britain

Recruiting

A Study of Implantation of Retinal Pigment Epithelium in Subjects with Acute Wet Age-Related Macular Degeneration

ClinicalTrials.gov ID NCT01691261

Sponsor Moorfields Eye Hospital NHS Foundation Trust
Information provided by Moorfields Eye Hospital NHS Foundation Trust (Responsible Party)
Last Update Posted 2022-08-22

Brief Summary
Phase 1 trial of retinal pigment epithelium replacement in subjects with wet age-related macular degeneration in whom there is rapidly progressing vision loss.

Detailed Description
Phase 1, open-label, safety, and feasibility study of implantation of PF-05206388 (human embryonic stem cell-derived retinal pigment epithelium) in subjects with wet age-related macular degeneration and rapid vision loss.

Official Title
Phase 1, Open-label, Safety and Feasibility Study of Implantation of Pf-05206388 (Human Embryonic Stem Cell-Derived Retinal Pigment Epithelium (RPE) Living Tissue Equivalent) in Subjects With Acute Wet Age-Related Macular Degeneration and Recent Rapid Vision Decline

Conditions
Age-Related Macular Degeneration

Study Type
Interventional

Enrollment (Estimated)
10

Intervention/Treatment
Biological: PF-05206388

Phase
Phase 1

Other Study ID Numbers
B4711001
2011-005493-37 (EudraCT Number)

Study Start (Actual)
2021-10-14

Primary Completion (Estimated)
2023-10-10

Study Completion (Estimated)
2024-03-14

Study Contact
Name: Tania West
Phone Number: 02072533411 ext 2036
Email: moorfields.resadmin@nhs.net

Study Contact Backup
Name: Daniela Narvaez
Phone Number: 02072533411 ext 2036
Email: moorfields.resadmin@nhs.net
United Kingdom

London, United Kingdom, EC1V 2PD
Recruiting
Moorfields Eye Hospital NHS Foundation Trust

London, United Kingdom, EC1V 2PD
Recruiting
Moorfields Eye Hospital NHS Foundation Trust

Contact
Moorfields R&D
020 7253 3411 ext 2937moorfields.resadmin@nhs.net

Eligibility Criteria
Ages Eligible for Study
60 years and older (adult, older adult)

Accepts Healthy Volunteer
No

Sexes Eligible for Study
All

Description

Inclusion Criteria

- Male and/or post-menopausal female subjects aged 60 years or above.
- Diagnosis of wet age-related macular degeneration (AMD) plus rapid recent vision decline.
- An informed consent document signed and dated by the subject or a legal representative.

Exclusion Criteria

- Pregnant females; breastfeeding females; and females of childbearing potential.
- Treatment with an investigational drug within 30 days (or as determined by the local requirement, whichever is longer) or 5 half-lives preceding the first dose of study medication.
- Current or previous significant other ocular disease in the study eye, as determined by the investigator.

Design Details
Primary Purpose: Treatment
Allocation: N/A
Interventional model: single-group assignment
Masking: none (open label)

Arms and interventions

Participant group/arm	Intervention/treatment
Experimental: Treatment PF-05206388 Retinal Pigment Epithelium living tissue equivalent for intraocular use in the form of a monolayer of retinal pigmented epithelial (RPE) cells immobilized on a polyester membrane	Biological: PF-05206388 • PF-05206388 will be provided as a retinal pigment epithelium living tissue equivalent for intraocular use in the form of a monolayer of retinal pigmented epithelial (RPE) cells immobilized on a polyester membrane. The membrane is approximately 6 × 3 mm and will contain a confluent layer of RPE cells, at a nominal dose of 17 mm^2. The implant is intended to be lifelong

Primary outcome measures

Outcome measure	Measure description	Time frame
Incidence and severity of adverse events.	The number of adverse events (AE) and serious adverse events (SAE) noted during the study and an assessment of whether they are trial product related	52 weeks
Change in baseline in ETDRS best-corrected visual acuity (BCVA)—Proportion of subjects with an improvement of 15 letters or more at Week 24	The number of patients with a difference between the baseline BCVA and BCVA at 24 weeks in ETDRS letters, where the differences is 15 letters or more, as a percentage of the total number of cases	24 weeks

Secondary outcome measures

Outcome measure	Measure description	Time frame
Change in baseline in ETDRS best-corrected visual acuity (BCVA)—proportion of subjects with an improvement of 15 letters or more	The number of patients with a difference between the baseline BCVA and BCVA at weeks 1, 2, 4, 8, 12, 16, 36, 52 in ETDRS letters, where the differences are 15 letters or more, as a percentage of the total number of cases	Weeks 1, 2, 4, 8, 12, 16, 36, 52
Mean change of best-corrected visual acuity (BCVA) from baseline by study visit	The mean difference between the baseline BCVA and final BCVA in ETDRS letters for all cases	52 weeks
Position of PF-05206388 by serial biomicroscopic evaluation	Measurement of movement in millimeters and rotation in degrees measure relative to baseline, at day 2 and weeks 1, 2, 4, 8, 10, 12, 16, 24, 36, and 52	Day 2 and weeks 1, 2, 4, 8, 10, 12, 16, 24, 36, 52
Position and presence of pigmented RPE cells by serial fundus photography	The subjective reporting of area of pigmentation as a % with cross-reference to the OCT at weeks 2, 4, 8, 10, 12, 16, 24, 36, and 52	Weeks 2, 4, 8, 10, 12, 16, 24, 36, 52
Mean change from baseline in contrast sensitivity by Pelli Robson test	The mean difference between the baseline BCVA and final BCVA in Pelli Robson letters read, across all subjects	Weeks 24, 52
Change in liver and renal function by blood tests and liver ultrasound	Record of any abnormalities in liver and renal function on blood testing and any abnormalities detected on the liver ultrasound	Weeks 24 and 52
Change in leakage or perfusion in normal fundal vasculature and presence of abnormal vasculature by fundus fluorescein angiography	Assessment and noting of abnormalities on fundus fluorescein angiography at weeks 4, 8, 12, 24, and 52	Weeks 4, 8, 12, 24 and 52
Change in central 30° of visual function by Humphrey field test	Recording and reporting of any changes on the central 30° field on the automated Humphrey field test at weeks 4, 8, 12, 24, and 52	Weeks 4, 8, 12, 24 and 52
Change in thickness of RPE layer by B-mode orbital ultrasound	Recording of any changes in thickness of RPE layer by B-mode orbital ultrasound carried out by the ocular oncologist or medical physicist at weeks 4, 8, 16, 24, 36, 52	Weeks 4, 8, 16, 24, 36, 52

Sponsor
Moorfields Eye Hospital NHS Foundation Trust

Collaborators
- University College, London

Investigators
- Study Director: Moorfields, Moorfields Eye Hospital NHS Foundation Trust

General Publications
No publications available.

Indonesia

Not Yet Recruiting

Role of UC-MSC and CM to Inhibit Vision Loss in Retinitis Pigmentosa Phase II/III

ClinicalTrials.gov ID NCT05909488

Sponsor PT. Prodia Stem Cell Indonesia
Information provided by PT. Prodia Stem Cell Indonesia (Responsible Party)
Last Update Posted 2023-06-18

Brief Summary
The study will perform UC-MSCs and CM transplantation. There are two groups with different dosages. The first group will be transplanted with 1.5 million cells; meanwhile, the second group is 5 million cells. Each group consists of 15 subjects. All groups will be transplanted via the peribulbar route. All groups will be observed until 6 months.

Detailed Description
The eyes we will give the transplant should be given an aseptic and antiseptic technique to prevent contamination from the inside and outside. Sterile cover attached to other parts of the face except for the eyes. 1.8 mL umbilical cord-derived mesenchymal stem cell (UC-MSC) preparations are suspended in conditioned medium (CM) until it reaches a 2 mL volume of cell suspension. Two dosages will be prepared, 1.5 million and 5 million of UC-MSC. Stem cell suspension will be injected by peribulbar, and patients will be given a quinolone antibiotic if the injection is done. On day 1st and day 7th after therapy, patients will be observed the presence of infection, inflammation, and increased eye pressure. On day 7th, day 30th, and day 90th after therapy, patients will do a visual field test, visual acuity test, electroretinography, funduscopy, and optical coherence tomography examination. The observations will be written in the table and analyzed by a statistic.

Study Contact
Name: Dr. Muhammad B Sasongko, Sp.M, M.Epid, PhD
Phone Number: +62811293476
Email: mb.sasongko@ugm.ac.id

Study Contact Backup
Name: Dr. Cynthia R Sartika, M.Si
Email: cynthia.sartika@prostem.co.id

Eligibility Criteria
Ages Eligible for Study
18–65 years (adult, older adult)

Accepts Healthy Volunteer
No
Sexes eligible for study
All

Description

Inclusion Criteria
- Vision more than 20/100
- Have more than 0.68 uV on conical receptor cell amplitude checked by ERG
- Visual field equivalent diameter more than 10o
- Willing to sign informed consent as research subjects
- Willing to do the peribulbar injection with mesenchymal stem cells isolated from umbilical cord tissue
- Willing to do visual field checks with Humphrey's perimetry, vision tests with Snellen boards, optical coherent tomography (OCT) examinations, electroretinogram examinations, and fill out a quality-of-life questionnaire

Exclusion Criteria
- Pregnant or nursing women
- Positive result of HIV test
- Have a history of eye tumors
- Under immunosuppressive treatment or other drugs that can affect the growth of transplanted stem cells
- Have another eye diseases such as diabetic retinopathy, uveitis, cataract, and glaucoma
- Do not come to control according to the schedule determined by the researcher (loss to follow-up)

Sponsor
PT. Prodia Stem Cell Indonesia

Collaborators
No information provided

Investigators
- Principal Investigator: Dr. Muhammad B Sasongko, Sp.M, M.Epid, PhD, Gadjah Mada University, Faculty of Medicine
- Study Chair: Dr. Cosmos O Mangunsong, Sp.M, Jakarta Eye Center
- Study Chair: Dr. Rifa Widyaningrum, M.Sc, PhD, Sardjito Hospital
- Study Chair: Rima Haifa, S.Si, Prodia StemCell Indonesia
- Study Chair: Marsya N Kirana, S.T., M.Sc, Prodia StemCell Indonesia

General Publications
No publications available.

Japan

Recruiting

Clinical Study of PAL-222 Targeting Patients with Myopic Chorioretinal Atrophy (PAMyCA)

ClinicalTrials.gov ID NCT05658237

Sponsor PharmaBio Corporation
Information provided by PharmaBio Corporation (Responsible Party)
Last Update Posted 2023-03-20

Brief Summary
The goal of this clinical trial is to assess efficacy and safety in patients with myopic chorioretinal atrophy.

The main questions it aims to answer are:

- Percentage of changes in the chorioretinal atrophic area.

Participants will be implanted one sheet of PAL-222 into the subretinal space through pars plana vitrectomy.

Researchers will compare non-therapeutic eye to see if the changes are significantly different.

Official Title
Phase I/IIa Clinical Study of PAL-222 Targeting Patients With Myopic Chorioretinal Atrophy (PAMyCA)

Conditions
Myopic Chorioretinal Atrophy

Study Type
Interventional

Enrollment (Estimated)
10

Intervention/Treatment
Procedure: Pars plana vitrectomy

Phase
Not Applicable

Other Study ID Numbers
PAL-222-P1

Study Start (Actual)
2023-02-17

Primary Completion (estimated)
2024-08-31

Study Completion (Estimated)
2024-10-31
Name: Toshihiko Sugioka
Phone Number: +81 44-589-6479
Email: clinicaltrial@pharmabio.co.jp
Japan
Nagoya, Aichi, Japan
Recruiting
Nagoya City University Hospital

Eligibility Criteria
Ages Eligible for Study
20 years and older (adult, older adult)

Accepts Healthy Volunteer
No

Sexes Eligible for Study
All

Description

Inclusion Criteria
- Patients aged 20 years or older at the time of consent acquisition
- Patients with binocular intensity myopia (myopia of −6.0 diopter (D) or more) or patients with axial length (26 mm) or more equivalent to −6.0 D
- Corrected characters of the test eye patients with visual acuity of less than 60 characters (equivalent to decimal visual acuity 0.32)
- Patients diagnosed with binocular myopic chorioretinal atrophy and having atrophy of 1 papillary diameter (1.5 mm) or more in the area within 2 papillary diameter (3.0 mm) including the fovea centralis in the subject eye
- Patients without active choroidal neovascularization

Exclusion Criteria
- Patients with abnormal findings that pose a problem in clinical trial participation in screening tests.
- Patients with positive hepatitis B surface (HBs) antigen, hepatitis C virus (HCV) antibody, human immunodeficiency virus (HIV) antibody, human T-lymphotropic virus type 1 (HTLV-1) antibody, syphilis serum reaction
- Patients with allergies to human serum albumin antibiotics, trypsin
- Patients with eye infections
- Patients with other retinal diseases (diabetic retinopathy, hypertensive retinopathy, vascular occlusion)
- Patients with confirmed optic nerve atrophy
- Patients with glaucoma who cannot control intraocular pressure

- Patients with findings associated with myopic traction macular disease (apparent posterior macular tumor, vitreous macular traction, retinal separation, macular hole detachment, macular hole)
- Patients with atrophy in all areas of the circle within 3 papilla diameter (4.5 mm) (9.0 mm in diameter) from the fovea
- Patients with corrected visual acuity of control eye 0.08 or less
- Patients with severe blood disorders, heart failure, liver disorders, and renal disorders
- Patients diagnosed with malignant tumor within 5 years or patients requiring treatment
- Pregnant women, lactating women, patients wishing to become pregnant during the trial period
- Patients who cannot discontinue anticoagulants or antiplatelet drugs before the trial
- Patients with drug addiction or alcoholism
- Patients receiving treatment with vascular endothelial growth factor inhibitor or photodynamic therapy within 6 months prior to transplantation
- Patients who underwent eye surgery (excluding lens reconstruction) within 3 months before transplantation. In the case of cataract surgery, if 7 days or more have passed since surgery and postoperative inflammation is stable, registration is possible

Design Details
Primary purpose: treatment
Allocation: N/A
Interventional model: single-group assignment
Masking: none (open label)

Arms and interventions

Participant group/arm	Intervention/treatment
Experimental: therapeutic group	Procedure: Pars plana vitrectomy • One sheet of PAL-222 is implanted into the subretinal space of either of eyes through pars plana vitrectomy

Primary outcome measures

Outcome measure	Measure description	Time frame
Percentage of changes in the chorioretinal atrophic area	Assessment by the principal investigator or sub-investigator will be conducted and central committee will adjudicate the assessment	Just before transplantation, after 3 days and 1, 2, 4, 8, 12, 16, 20, 24, 32, 40, 52 weeks

Sponsor
PharmaBio Corporation

Collaborators
No information provided

Investigators

- Study Director: Hitoshi Kusano, M.D., PhamaBio Corporation

General Publications

No publications available

Turkey

Recruiting

The Effect of Stem Cells and Stem Cell Exosomes on Visual Functions in Patients With Retinitis Pigmentosa

ClinicalTrials.gov ID NCT05413148

Sponsor TC Erciyes University
Information provided by Kuddusi Erkılıç, TC Erciyes University (Responsible Party)
Last Update Posted 2022-09-07

Brief Summary

This clinical trial aims to study the efficacy of umbilical cord-derived mesenchymal stem cells and their exosomes in the treatment of retinitis pigmentosa. The participants will be randomized into 3 groups. Functional and structural parameters will be compared before and after the injections and also will be compared among the groups to reveal whether stem cell and their exosomes are more effective than placebo.

Detailed Description

Retinitis pigmentosa (RP) is one of the leading causes of vision loss and irreversible blindness. It is stated that 1.4 million people are affected in the world and its prevalence is 1:4000. Stem cell applications are methods that are increasingly gaining importance in the field of regenerative medicine and ophthalmology. In addition to many experimental studies in retinal diseases including retinitis pigmentosa, there are also clinical studies reporting successful results on humans. The most widely used stem cell group in current clinical practice is mesenchymal stem cells. The advantages are that they are relatively easy to obtain and do not induce severe immune responses. Clinical studies in stem cell therapies for the eye suggest that stem cells benefit the surrounding tissue by secreting support such as growth factors, and extracellular vesicles, mostly due to their trophic and paracrine effects. In intraocular applications, there has been a tendency to apply mesenchymal stem cells around/outside the eye, including subtenon and suprachoroidal application due to side effects such as retinal detachment and epiretinal membrane.

Other cellular products that have been studied in animal studies and clinical research in recent years are exosomes. Exosomes are a subgroup of extracellular vesicles released from cells. Exosomes are extracellular vesicles 30–150 nm in size and carry at least one of the exosomal proteins CD63, CD9, CD81, syntenin-1, and TSG101. Exosomes are secreted from different cells in the organism and are also found in body fluids. Exosomes play a role in the transportation of biomolecules such as lipids, carbohydrates, nucleic acids, and proteins from one cell to another and in this way, they function in genetic information transfer, and reprogramming of the opposite cell and intercellular communication. One of the sources of exosomes is mesenchymal stem cells that secrete high amounts of exosomes. Mesenchymal stem cell-derived exosomes have been reported to have therapeutic effects, just like mesenchymal stem cells. It is thought that stem cells also exert their clinical effects through paracrine factors they secrete rather than being implanted in the tissue. However, the mechanism of action of mesenchymal stem cells has not been fully understood, and exosomes are thought to contribute to the effect. The storage and durability of exosomes may provide advantages in therapeutic applications compared to mesenchymal stem cells and may provide more immunological advantages in allogeneic applications compared to cellular treatments since they do not contain cells.

One of the sources from which mesenchymal stem cells can be obtained is Wharton jelly, which is the mesenchymal tissue of the umbilical cord. Stem cells originating from Wharton gel are a good source of stem cells due to their high differentiation capacity, high regeneration capacity, constant doubling time, high proliferation capacity, low immunogenicity, and no ethical problems due to non-invasive methods, and easy availability.

This clinical trial will enroll patients diagnosed with retinitis pigmentosa. The diagnosis of RP is made by clinical fundus examination, examinations in our clinic, and genetic analysis.

At the beginning of the study, all volunteers will undergo a full ophthalmologic examination including best-corrected visual acuity, anterior segment and fundus examination with a slit lamp, OCT (optical coherence tomography), VF (visual field), ERG (electroretinogram), mfERG (multifocal ERG), OCTA (optical coherence tomography). Retinal and choroidal thickness and ellipsoid bandwidth in OCT, MD (mean deviation) value showing sensitivity in VF, amplitude and latency times in ERG and mf-ERG, vascular density, and vascular flow ratios in OCTA will be recorded.

The volunteers will be randomized into three. The 1st group will be the placebo group (45 volunteers) and 0.5 cc saline physiological saline will be applied to the subtenon space. Second group of patients (45 volunteers) will undergo a subtenon injection of Wharton gel-derived mesenchymal stem cell suspension. A suspension containing mesenchymal stem cell exosomes from Wharton jelly will be applied to the subtenon distance to the 3rd group of patients (45 volunteers). A single eye of each volunteer will be included. Informed consent will be obtained from the volunteers.

Allogeneic human Wharton gel tissue-derived mesenchymal stem cells and their exosomes will be obtained from healthy female volunteers, who is over 18 years old, under sterile conditions after HBV, HCV, HIV virus, and "Treponema Pallidum" (VDRL) tests were performed. Cord tissue obtained from birth will be carried in a sterile transfer solution and will be processed to obtain stem cells in Erciyes University Genome and Stem Cell GMP (good manufacturing practices) certified laboratories. Exosomes will be obtained by the ultracentrifugation method in the same facility.

For injection into the subtenon distance, the conjunctiva and tenon will be opened with a small incision about 10 mm away from the lower temporal area, and then, 20 gauge curved subtenon cannula will be advanced parallel to the sclera and injected into the posterior subtenon area. All volunteers will be prescribed antibiotics and steroid drops after the surgical procedure.

Full ophthalmological examination and examinations at the beginning of the study will be performed and recorded again on the 1st day, 1st month, 2nd month, 3rd month, and 6th months of the anterior segment and fundus examination, OCT (optical coherence tomography), VF (visual field), ERG (electroretinogram), mfERG (multifocal ERG), OCTA (optical coherence tomography) examinations will be performed.

In addition, a questionnaire evaluating the visual functions subjectively will be administered at the 1st-, 3rd-, and 6th-month visits. (National Eye Institute Visual Functioning Questionnaire—25/National Eye Institute (USA) Visual Function Questionnaire). Two masked investigators will perform analyses.

Post-procedure 1st-day, 1st-month, 2nd-month, 3rd-month, and 6th-month data will be compared among study groups and within groups.

Official Title

The Effect of Wharton Jelly-derived Mesenchymal Stem Cells and Stem Cell Exosomes on Visual Functions in Patients With Retinitis Pigmentosa

Conditions

Retinitis Pigmentosa

Study Type

Interventional

Enrollment (Estimated)

135

Intervention/Treatment

Biological: Subtenon injection of Wharton jelly-derived mesenchymal stem cells
Biological: Subtenon injection of Wharton jelly-derived mesenchymal stem cell exosomes
Other: Placebo

Phase

Phase 2 and 3

Other Study ID Numbers

TSG-2021-11599

Study Start (Actual)

2022-08-05

Primary Completion (Estimated)
2023-07-01

Study Completion (Estimated)
2023-12-15

Study Contact
Name: Osman Ahmet POLAT, MD
Phone Number: 905424395196
Email: osmanahmet@gmail.com
Turkey

Kayseri, Turkey, 38039
Recruiting
Erciyes University, Faculty of Medicine

Contact
Osman Ahmet Polat, MD
+905424395196 osmanpolat@erciyes.edu.tr

Eligibility Criteria
Ages Eligible For Study
18–70 years (adult, older adult)

Accepts Healthy Volunteer
No

Sexes Eligible for Study
All

Description

Inclusion Criteria
- 18 years and over
- Diagnosis of retinitis pigmentosa: by clinical history, fundus examination, visual field (GA), electroretinogram (ERG), and genetic analysis
- Visual field loss
- The best-corrected visual acuity of 0.05 on the Snellen chart
- The MD (mean deviation) value in the visual field is between 33.0 and −5.0 dB
- Intraocular pressure value below 22 mmHg

Exclusion Criteria
- Presence of cataracts or other media opacities that may affect imaging and tests
- Diagnosis of glaucoma
- History of ocular surgery or injection in the last 6 months
- Visual level too low for examinations (0.05 and below)
- Diagnosis of any systemic disease (such as diabetes, uncontrolled hypertension, neurological disease)
- Smoking and substance abuse

Design Details

Primary purpose: treatment
Allocation: randomized
Interventional model: parallel assignment
Masking: triple (participant investigator outcomes assessor)

Arms and interventions

Participant group/arm	Intervention/treatment
Active comparator: Wharton jelly-derived mesenchymal stem cell Single subtenon injection of Wharton jelly-derived mesenchymal stem cells will be performed on a single eye after randomization	Biological: Subtenon injection of Wharton jelly-derived mesenchymal stem cells • Single subtenon injection for single eye
Active comparator: Mesenchymal stem cell exosome (Wharton jelly-derived) Single subtenon injection of mesenchymal stem cell exosomes (Wharton jelly-derived) will be performed on a single eye after randomization	Biological: Subtenon injection of Wharton jelly-derived mesenchymal stem cell exosomes • Single subtenon injection for single eye
Placebo comparator: Placebo A single subtenon injection of saline will be performed on a single eye after randomization	Other: Placebo • Single subtenon injection of saline for a single eye

Primary outcome measures

Outcome measure	Measure description	Time frame
Evaluation of best-corrected visual acuity change	Visual acuity change with LogMar Chart	Up to 6 months
Evaluation of visual field change	Mean deviation values of automatic visual field testing	Up to 6 months

Secondary outcome measures

Outcome measure	Measure description	Time frame
Evaluation of multifocal ERG changes	Amplitudes changes in multifocal ERG testing	Up to 6 months
Evaluation of ERG changes	Amplitude changes in multifocal ERG testing	Up to 6 months
Evaluation of optical coherence tomography changes	Retinal thickness changes in optical coherence tomography change	Up to 6 months
Evaluation of optical coherence tomography angiography changes	Vascular density changes in optical coherence tomography angiography	Up to 6 months

Sponsor

TC Erciyes University

Collaborators

No information provided

Investigators

- Principal Investigator: Kuddusi ERKILIÇ, Professor, Erciyes University Medical Faculty

General Publications

- Limoli PG, Limoli CSS, Morales MU, Vingolo EM. Mesenchymal stem cell surgery, rescue and regeneration in retinitis pigmentosa: clinical and rehabilitative prognostic aspects. Restor Neurol Neurosci. 2020;38(3):223–237. https://doi.org/10.3233/RNN-190970.
- Bhattacharya S, Gangaraju R, Chaum E. Recent Advances in Retinal Stem Cell Therapy. Curr Mol Biol Rep. 2017 Sep;3(3):172–182. https://doi.org/10.1007/s40610-017-0069-3. Epub 2017 Jul 10.
- Limoli PG, Vingolo EM, Limoli C, Nebbioso M. Stem Cell Surgery and Growth Factors in Retinitis Pigmentosa Patients: Pilot Study after Literature Review. Biomedicines. 2019;7(4):94. https://doi.org/10.3390/biomedicines7040094.
- Borkowska-Kuczkowska A, Slugocka D, Swiatkowska-Flis B, Boruczkowski D. The use of mesenchymal stem cells for the treatment of progressive retinal diseases: a review. Regen Med. 2019;14(4):321–329. https://doi.org/10.2217/rme-2019-0022. Epub 2019 Apr 12.
- Jin ZB, Gao ML, Deng WL, Wu KC, Sugita S, Mandai M, Takahashi M. Stemming retinal regeneration with pluripotent stem cells. Prog Retin Eye Res. 2019;69:38–56. https://doi.org/10.1016/j.preteyeres.2018.11.003. Epub 2018 Nov 9.
- Ozmert E, Arslan U. Management of retinitis pigmentosa by Wharton's jelly derived mesenchymal stem cells: preliminary clinical results. Stem Cell Res Ther. 2020;11(1):25. https://doi.org/10.1186/s13287-020-1549-6.
- Oner A. Stem cell treatment in retinal diseases: recent developments. Turk J Ophthalmol. 2018;48(1):33–38. https://doi.org/10.4274/tjo.89972. Epub 2018 Feb 23.
- Cotrim CC, Jorge R, Oliveira MC, Pieroni F, Messias AMV, Siqueira RC. Clinical studies using stem cells for treatment of retinal diseases: state of the art. Arq Bras Oftalmol. 2020;83(2):160–167. https://doi.org/10.5935/0004-2749.20200037.
- Huo DM, Dong FT, Yu WH, Gao F. Differentiation of mesenchymal stem cell in the microenvironment of retinitis pigmentosa. Int J Ophthalmol. 2010;3(3):216–9. https://doi.org/10.3980/j.issn.2222-3959.2010.03.08. Epub 2010 Sep 18.
- Satarian L, Nourinia R, Safi S, Kanavi MR, Jarughi N, Daftarian N, Arab L, Aghdami N, Ahmadieh H, Baharvand H. Intravitreal injection of bone marrow mesenchymal stem cells in patients with advanced retinitis pigmentosa; a safety study. J Ophthalmic Vis Res. 2017;12(1):58–64. https://doi.org/10.4103/2008-322X.200164.
- Kuriyan AE, Albini TA, Townsend JH, Rodriguez M, Pandya HK, Leonard RE, Parrott MB, Rosenfeld PJ, Flynn HW Jr, Goldberg JL. Vision loss after intravitreal injection of autologous "stem cells" for AMD. N Engl J Med. 2017;376(11):1047–1053. https://doi.org/10.1056/NEJMoa1609583.
- Harrell CR, Fellabaum C, Jovicic N, Djonov V, Arsenijevic N, Volarevic V. Molecular mechanisms responsible for therapeutic potential of mesenchymal

stem cell-derived secretome. Cells. 2019;8(5):467. https://doi.org/10.3390/cells8050467.

- Klingeborn M, Dismuke WM, Bowes Rickman C, Stamer WD. Roles of exosomes in the normal and diseased eye. Prog Retin Eye Res. 2017;59:158–177. https://doi.org/10.1016/j.preteyeres.2017.04.004. Epub 2017 Apr 29.
- Harrell CR, Simovic Markovic B, Fellabaum C, Arsenijevic A, Djonov V, Arsenijevic N, Volarevic V. Therapeutic potential of mesenchymal stem cell-derived exosomes in the treatment of eye diseases. Adv Exp Med Biol. 2018;1089:47–57. https://doi.org/10.1007/5584_2018_219.
- Mead B, Tomarev S. Extracellular vesicle therapy for retinal diseases. Prog Retin Eye Res. 2020;79:100849. https://doi.org/10.1016/j.preteyeres.2020.100849. Epub 2020 Mar 10.
- Yaghoubi Y, Movassaghpour A, Zamani M, Talebi M, Mehdizadeh A, Yousefi M. Human umbilical cord mesenchymal stem cells derived-exosomes in diseases treatment. Life Sci. 2019;233:116733. https://doi.org/10.1016/j.lfs.2019.116733. Epub 2019 Aug 5.
- Zhang X, Liu J, Yu B, Ma F, Ren X, Li X. Effects of mesenchymal stem cells and their exosomes on the healing of large and refractory macular holes. Graefes Arch Clin Exp Ophthalmol. 2018;256(11):2041–2052. https://doi.org/10.1007/s00417-018-4097-3. Epub 2018 Aug 30.
- Ma M, Li B, Zhang M, Zhou L, Yang F, Ma F, Shao H, Li Q, Li X, Zhang X. Therapeutic effects of mesenchymal stem cell-derived exosomes on retinal detachment. Exp Eye Res. 2020;191:107899. https://doi.org/10.1016/j.exer.2019.107899. Epub 2019 Dec 19.
- Yu B, Shao H, Su C, Jiang Y, Chen X, Bai L, Zhang Y, Li Q, Zhang X, Li X. Exosomes derived from MSCs ameliorate retinal laser injury partially by inhibition of MCP-1. Sci Rep. 2016;6:34562. https://doi.org/10.1038/srep34562.
- Zhang W, Wang Y, Kong Y. Exosomes derived from mesenchymal stem cells modulate mir-126 to ameliorate hyperglycemia-induced retinal inflammation via targeting HMGB1. Invest Ophthalmol Vis Sci. 2019;60(1):294–303. https://doi.org/10.1167/iovs.18-25617.
- Hajrasouliha AR, Jiang G, Lu Q, Lu H, Kaplan HJ, Zhang HG, Shao H. Exosomes from retinal astrocytes contain antiangiogenic components that inhibit laser-induced choroidal neovascularization. J Biol Chem. 2013;288(39):28058–67. https://doi.org/10.1074/jbc.M113.470765. Epub 2013 Aug 7.
- Ha DH, Kim SD, Lee J, Kwon HH, Park GH, Yang SH, Jung JY, Lee JH, Park SR, Youn J, Lee SH, Kim JE, Lim J, Lee HK, Cho BS, Yi YW. Toxicological evaluation of exosomes derived from human adipose tissue-derived mesenchymal stem/stromal cells. Regul Toxicol Pharmacol. 2020;115:104686. https://doi.org/10.1016/j.yrtph.2020.104686. Epub 2020 May 22.
- Pan D, Chang X, Xu M, Zhang M, Zhang S, Wang Y, Luo X, Xu J, Yang X, Sun X. UMSC-derived exosomes promote retinal ganglion cells survival in a rat model of optic nerve crush. J Chem Neuroanat. 2019;96:134–139. https://doi.org/10.1016/j.jchemneu.2019.01.006. Epub 2019 Jan 10.

- Zhao T, Liang Q, Meng X, Duan P, Wang F, Li S, Liu Y, Yin ZQ. Intravenous infusion of umbilical cord mesenchymal stem cells maintains and partially improves visual function in patients with advanced retinitis pigmentosa. Stem Cells Dev. 2020;29(16):1029–1037. https://doi.org/10.1089/scd.2020.0037. Epub 2020 Jul 15.

United States: Anterior Segment

Not Yet Recruiting

Efficacy of Locally Delivered Allogeneic Mesenchymal Stromal Cells

ClinicalTrials.gov ID NCT05705024

Sponsor University of Illinois at Chicago
Information provided by Ali R Djalilian, University of Illinois at Chicago (Responsible Party)
Last Update Posted 2023-01-30

Brief Summary
The proposed Conventional Cohort Expansion Study involves the use of Mesenchymal Stromal Cells (MSCs) derived from the bone marrow. We previously studied the safety of subconjunctival injection of allogeneic bone marrow-derived MSCs in patients with nonhealing epitheliopathy (IRB Protocol 2020-0334). In the present study, we want to study the efficacy of this treatment on chronic epitheliopathies.

Detailed Description
The "Efficacy of Locally Delivered Allogeneic Mesenchymal Stem Cells for Promoting Corneal Repair Study" otherwise known as the "MSC Study" is designed to assess the safety of allogeneic bone marrow-derived MSC secreted factor on the ocular surface via subconjunctival injection of MSC and also obtain a preliminary observation on the following:

- Epithelial barrier integrity and/or wound closure.
- Development of scarring.
- Final visual acuity.

The objective is to improve clinical outcomes in significant nonhealing corneal wounds. To achieve these goals, the MSC study will include a phase II efficacy study.

Official Title
Efficacy of Locally Delivered Allogeneic Mesenchymal Stem Cells for Promoting Corneal Repair

Conditions
Corneal ulcer

Study Type
Interventional

Enrollment (Estimated)
38

Intervention/Treatment
Biological: mesenchymal stromal cells
Other: Control Solution

Phase
Phase 2

Other Study ID Numbers
2022-0751
W81XWH-18-1-0661 (Other Grant/Funding Number) (OTHER_GRANT: United States Department of Defense)

Study Start (Estimated)
2023-05-01

Primary Completion (Estimated)
2024-09-28

Study Completion (Estimated)
2024-09-28

Study Contact
Name: Ali R Djalilian, MD
Phone Number: 312-996-8937
Email: adjalili@uic.edu

Study Contact Backup
Name: Charlotte E Joslin, OD, PhD
Phone Number: 312-996-5410
Email: charjosl@uic.edu

United States
Chicago, Illinois, United States, 60612
Department of Ophthalmology and Visual Sciences

Contact
Ali R Djalilian, MD
312-996-8937 adjalili@uic.edu

Contact
Charlotte E Joslin, OD, PhD
312-996-5410 charjosl@uic.edu

Eligibility Criteria
Ages Eligible for Study
18 years and older (adult, older adult)

Accepts Healthy Volunteer
No

Sexes Eligible for Study
All

Description

Inclusion Criteria
Visual Acuity

- Best-corrected distance visual acuity (BCDVA) score ≤75 ETDRS letters, (≥0.2 LogMAR, ≤20/32 Snellen or ≤0.625 decimal fraction) in the affected eye.

Ocular Health
- Patients with non-resolving corneal epitheliopathy or epithelial defect after two or more weeks of standard non-surgical treatments (e.g., preservative-free artificial tears, gels, or ointments; discontinuation of preserved topical drops; anti-inflammatory therapy, soft bandage contact lens).
- No objective clinical evidence of improvement in the last 2 weeks (≤50% reduction in fluorescein staining or ≤50% reduction in longest diameter of the epithelial defect).
- If both eyes have chronic epithelial disease, the eye with the worse epithelial disease will be treated.
- Evidence of impaired epithelial barrier manifested by fluorescein staining of the epithelium with a score 10 or higher by National Eye Institute grading.
- Patients with stage 1 (no epithelial defect), stage 2 (persistent epithelial defect, PED; without stromal loss), or stage 3 (corneal ulcer; with stromal loss) neurotrophic keratopathy 25–27 limited to ≤80% corneal diameter.

Study Procedures
- Only patients who satisfy all informed consent requirements may be included in the study. The patient and/or his/her legal representative must read, sign and date the informed consent document before any study-related procedures are performed. The informed consent form signed by patients and/or legal representatives must have been approved by the IRB for the current study.
- Patients must have the ability and willingness to comply with study procedures.

Exclusion Criteria
Visual Acuity

- Best-corrected distance visual acuity (BCDVA) score better than 75 ETDRS letters, or 0.2 LogMAR, or 20/32 Snellen or 0.625 decimal fraction in the affected eye.

Ocular Health

- Ocular drug toxicity less than 2 weeks ago.
- Any active ocular infection (bacterial, viral, fungal, or protozoal) or active ocular inflammation in the affected eye.
- History of any ocular surgery (including laser or refractive surgical procedures) in the affected eye within the 3 months before study enrollment. (An exception to the preceding statement will be allowed if the ocular surgery is considered to be the cause of the PED. Ocular surgery in the affected eye will not be allowed during the study treatment period and elective ocular surgery procedures should not be planned during the duration of the follow-up period unless the patient will be involved in corneal thinning of more than one-third corneal stroma, corneal melting or perforation.
- Prior surgical procedure(s) for the treatment of a chronic corneal epitheliopathy (e.g., complete tarsorrhaphy and conjunctival flap) in the affected eye with the exception of amniotic membrane transplantation. Patients previously treated with amniotic membrane transplantation may only be enrolled 2 weeks after the membrane has disappeared within the area of the chronic corneal epitheliopathy or corneal ulcer or at least 6 weeks after the date of the amniotic membrane transplantation procedure. Patients previously treated with Botox (botulinum toxin) injections used to induce pharmacologic blepharoptosis are eligible for enrollment only if the last injection was given at least 90 days prior to enrollment in the study.
- Chronic corneal epitheliopathy in the background of endothelial decompensation that needs corneal graft.
- Anticipated need for punctual occlusion during the study treatment period. Patients with punctual occlusion or punctual plugs inserted prior to the study are eligible for enrollment provided that the punctual occlusion is maintained during the study.
- Evidence of corneal ulceration involving the posterior third of the corneal stroma, corneal melting, or perforation in the affected eye.
- Presence or history of any ocular or systemic disorder or condition that might hinder the efficacy of the study treatment or its evaluation, could possibly interfere with the interpretation of study results, or could be judged by the investigator to be incompatible with the study visit schedule or conduct (e.g., progressive or degenerative corneal or retinal conditions, uveitis, optic neuritis, poorly controlled diabetes, autoimmune disease, systemic infection, neoplastic diseases).
- Patients with uncontrolled eyelid abnormality that preclude appropriate eyelid closure or including eyelash abnormality.

Study Procedures

- Known hypersensitivity to one of the components of the study or procedural medications (e.g., fluorescein).
- History of drug, medication, or alcohol abuse or addiction.
- Use of any investigational agent within 4 weeks of screening visit.
- Participation in another clinical study at the same time as the present study.

- Participants who are pregnant at the time of study enrollment will be excluded; pregnancy is identified according to the patient's self-report/positive βhCG.

Design Details
Primary Purpose: Treatment
Allocation: Randomized
Interventional Model: Parallel Assignment
Masking: Double (Participant Care Provider)

Masking Description: Thirty-eight (38) patients will be recruited in this randomized clinical trial study with a 2:1 allocation for randomization. Twenty-five (25) patients will be recruited for the MSC treatment and thirteen (13) patients will be assigned to the control group. At any stage, if the results of MSC group were significantly better than the control group, using the control group would be stopped. This process will be supervised by Charlotte Joselin who is the person in charge for blinding and for the formulation and distribution of the proper study drug for injection.

Arms and interventions

Participant group/arm	Intervention/treatment
Active comparator: Medium dose of allogenic MSC drops Dose of allogeneic MSC subconjunctival injection will be assigned 3,000,000 cells/150 μL	Biological: Mesenchymal stromal cells • Subconjunctival injection of allogeneic mesenchymal stromal cells masked clinical trial, patients with non-resolving corneal epithelial disease (i.e., refractory to standard treatments for at least 2 weeks) will receive a single subconjunctival injection of bone marrow-derived allogeneic MSCs or vehicle (CS5 freezing media, BioLife Solutions Inc, Bothell, WA), with continued follow-up for up to 90 days
Sham Comparator: Control Group For the control group, 50 μL of the freezing media (vehicle) will be injected	Other: Control Solution • For the control group, 50 μL of the freezing media (vehicle) will be injected

Primary outcome measures

Outcome measure	Measure description	Time frame
Improvement of corneal epithelial barrier and/or integrity (efficacy rate)	The proportion of patients with improved epithelial barrier and/or integrity from baseline to day 28 as determined by the investigator on slit lamp examination • Improved epithelial barrier, defined as a 50 % improvement in corneal fluorescein staining score • Improved epithelial integrity, defined as a healed epithelial defect	Day 28

Secondary outcome measures

Outcome measure	Measure description	Time frame
Visual acuity	Percent change in best-corrected distance visual acuity from baseline to day 90, as measured using standard ETDRS protocols	Baseline, days 1–7, 28, 60, 90
Corneal staining and NEI grading	Grading of fluorescein staining of the cornea	Baseline, days 1–7, 28, 60, 90
Ocular surface parameters	Changes in tear breakup time (TBUT), ocular surface disease index (OSDI), Lissamine green staining, and anesthetic Schirmer's test from baseline to day 28 and day 90	Baseline, day 28, 90
Corneal epithelial thickness	Percent change in corneal epithelial thickness from baseline to day 28 and day 90, as measured by anterior segment OCT (AS-OCT)	Day 28, 90
Patient symptoms	Changes in ocular discomfort visual analog scale (VAS) 0–100, where 0 is no discomfort and 100 the worst discomfort, from baseline to day 90	Baseline, days 1–7, 28, 60, 90
Time to epithelial healing	Time of improvement of epitheliopathy	Baseline, days 1–7, 28, 60, 90
Corneal scar	Change in the size of corneal scar (if present) from baseline to day 90, as documented by slit lamp photographs	Baseline, days 1–7, 28, 60, 90
Corneal neo-vascularization	Change in corneal vascularization on slit lamp photographs from baseline to day 90	Baseline, days 1–7, 28, 60, 90
Conjunctival injection	Change in conjunctival injection on slit lamp examination from baseline to day 90	Baseline, days 1–7, 28, 60, 90

Sponsor

University of Illinois at Chicago

Collaborators

- United States Department of Defense

Investigators

- Principal Investigator: Ali R Djalilian, MD, University of Illinois at Chicago
- Principal Investigator: Charlotte E Joslin, OD, PhD, University of Illinois at Chicago

General Publications

- Cockerham GC, Lemke S, Rice TA, Wang G, Glynn-Milley C, Zumhagen L, Cockerham KP. Closed-globe injuries of the ocular surface associated with combat blast exposure. Ophthalmology. 2014;121(11):2165–72. https://doi.org/10.1016/j.ophtha.2014.06.009. Epub 2014 Aug 11.
- Cockerham GC, Lemke S, Glynn-Milley C, Zumhagen L, Cockerham KP. Visual performance and the ocular surface in traumatic brain injury. Ocul Surf. 2013;11(1):25–34. https://doi.org/10.1016/j.jtos.2012.09.004. Epub 2012 Oct 5.
- Cockerham GC, Rice TA, Hewes EH, Cockerham KP, Lemke S, Wang G, Lin RC, Glynn-Milley C, Zumhagen L. Closed-eye ocular injuries in the Iraq and Afghanistan wars. N Engl J Med. 2011;364(22):2172–3. https://doi.org/10.1056/NEJMc1010683. No abstract available.
- Baradaran-Rafii A, Eslani M, Tseng SC. Sulfur mustard-induced ocular surface disorders. Ocul Surf. 2011;9(3):163–78. https://doi.org/10.1016/s1542-0124(11)70026-x.
- Prockop DJ, Oh JY. Mesenchymal stem/stromal cells (MSCs): role as guardians of inflammation. Mol Ther. 2012;20(1):14–20. https://doi.org/10.1038/mt.2011.211. Epub 2011 Oct 18.
- Mittal SK, Omoto M, Amouzegar A, Sahu A, Rezazadeh A, Katikireddy KR, Shah DI, Sahu SK, Chauhan SK. Restoration of corneal transparency by mesenchymal stem cells. Stem Cell Reports. 2016;7(4):583–590. https://doi.org/10.1016/j.stemcr.2016.09.001. Epub 2016 Sep 29.
- Wang LT, Ting CH, Yen ML, Liu KJ, Sytwu HK, Wu KK, Yen BL. Human mesenchymal stem cells (MSCs) for treatment towards immune-and inflammation-mediated diseases: review of current clinical trials. J Biomed Sci. 2016;23(1):76. https://doi.org/10.1186/s12929-016-0289-5.
- Yun YI, Park SY, Lee HJ, Ko JH, Kim MK, Wee WR, Reger RL, Gregory CA, Choi H, Fulcher SF, Prockop DJ, Oh JY. Comparison of the anti-inflammatory effects of induced pluripotent stem cell-derived and bone marrow-derived mesenchymal stromal cells in a murine model of corneal injury. Cytotherapy. 2017;19(1):28–35. https://doi.org/10.1016/j.jcyt.2016.10.007. Epub 2016 Nov 10.
- Ye J, Yao K, Kim JC. Mesenchymal stem cell transplantation in a rabbit corneal alkali burn model: engraftment and involvement in wound healing. Eye. 2006;20(4):482–90. https://doi.org/10.1038/sj.eye.6701913.
- Yao L, Li ZR, Su WR, Li YP, Lin ML, Zhang WX, Liu Y, Wan Q, Liang D. Role of mesenchymal stem cells on cornea wound healing induced by acute alkali burn. PLoS One. 2012;7(2):e30842. https://doi.org/10.1371/journal.pone.0030842. Epub 2012 Feb 17.
- Roddy GW, Oh JY, Lee RH, Bartosh TJ, Ylostalo J, Coble K, Rosa RH Jr, Prockop DJ. Action at a distance: systemically administered adult stem/progenitor cells (MSCs) reduce inflammatory damage to the cornea without engraftment and primarily by secretion of TNF-alpha stimulated gene/protein 6. Stem Cells. 2011;29(10):1572–9. https://doi.org/10.1002/stem.708.

- Oh JY, Kim MK, Shin MS, Lee HJ, Ko JH, Wee WR, Lee JH. The anti-inflammatory and anti-angiogenic role of mesenchymal stem cells in corneal wound healing following chemical injury. Stem Cells. 2008;26(4):1047–55. https://doi.org/10.1634/stemcells.2007-0737. Epub 2008 Jan 10.
- Ma Y, Xu Y, Xiao Z, Yang W, Zhang C, Song E, Du Y, Li L. Reconstruction of chemically burned rat corneal surface by bone marrow-derived human mesenchymal stem cells. Stem Cells. 2006;24(2):315–21. https://doi.org/10.1634/stemcells.2005-0046. Epub 2005 Aug 18.
- Li F, Zhao SZ. Control of cross talk between angiogenesis and inflammation by mesenchymal stem cells for the treatment of ocular surface diseases. Stem Cells Int. 2016;2016:7961816. https://doi.org/10.1155/2016/7961816. Epub 2016 Mar 24.
- Cejkova J, Trosan P, Cejka C, Lencova A, Zajicova A, Javorkova E, Kubinova S, Sykova E, Holan V. Suppression of alkali-induced oxidative injury in the cornea by mesenchymal stem cells growing on nanofiber scaffolds and transferred onto the damaged corneal surface. Exp Eye Res. 2013;116:312–23. https://doi.org/10.1016/j.exer.2013.10.002. Epub 2013 Oct 18.
- Putra I, Shen X, Anwar KN, Rabiee B, Samaeekia R, Almazyad E, Giri P, Jabbehdari S, Hayat MR, Elhusseiny AM, Ghassemi M, Mahmud N, Edward DP, Joslin CE, Rosenblatt MI, Dana R, Eslani M, Hematti P, Djalilian AR. Preclinical evaluation of the safety and efficacy of cryopreserved bone marrow mesenchymal stromal cells for corneal repair. Transl Vis Sci Technol. 2021;10(10):3. https://doi.org/10.1167/tvst.10.10.3.
- Jabbehdari S, Yazdanpanah G, Kanu LN, Anwar KN, Shen X, Rabiee B, Putra I, Eslani M, Rosenblatt MI, Hematti P, Djalilian AR. Reproducible derivation and expansion of corneal mesenchymal stromal cells for therapeutic applications. Transl Vis Sci Technol. 2020;9(3):26. https://doi.org/10.1167/tvst.9.3.26.
- Bartlett RS, Guille JT, Chen X, Christensen MB, Wang SF, Thibeault SL. Mesenchymal stromal cell injection promotes vocal fold scar repair without long-term engraftment. Cytotherapy. 2016;18(10):1284–96. https://doi.org/10.1016/j.jcyt.2016.07.005.
- Bara JJ, Richards RG, Alini M, Stoddart MJ. Concise review: Bone marrow-derived mesenchymal stem cells change phenotype following in vitro culture: implications for basic research and the clinic. Stem Cells. 2014;32(7):1713–23. https://doi.org/10.1002/stem.1649.
- Cook N, Hansen AR, Siu LL, Abdul Razak AR. Early phase clinical trials to identify optimal dosing and safety. Mol Oncol. 2015;9(5):997–1007. https://doi.org/10.1016/j.molonc.2014.07.025. Epub 2014 Aug 14.
- Al-Moujahed A, Chodosh J. Outcomes of an algorithmic approach to treating mild ocular alkali burns. JAMA Ophthalmol. 2015;133(10):1214–6. https://doi.org/10.1001/jamaophthalmol.2015.2302. No abstract available.
- Ghazaryan E, Zhang Y, He Y, Liu X, Li Y, Xie J, Su G. Mesenchymal stem cells in corneal neovascularization: Comparison of different application routes. Mol Med Rep. 2016;14(4):3104–12. https://doi.org/10.3892/mmr.2016.5621. Epub 2016 Aug 11.

- U.S. Department of Health and Human Services Food and Drug Administration Center for Biologics Evaluation and Research. Considerations for the Design of Early-Phase Clinical Trials of Cellular and Gene Therapy Products. Accessed September 26, 2017. https://www.fda.gov/downloads/BiologicsBloodVaccines/GuidanceComplianceRegulatoryInformation/Guidances/CellularandGeneTherapy/UCM564952.pdf.
- Sacchetti M, Lambiase A. Diagnosis and management of neurotrophic keratitis. Clin Ophthalmol. 2014;8:571–9. https://doi.org/10.2147/OPTH.S45921. eCollection 2014.
- Bonini S, Rama P, Olzi D, Lambiase A. Neurotrophic keratitis. Eye. 2003;17(8):989–95. https://doi.org/10.1038/sj.eye.6700616.
- Semeraro F, Forbice E, Romano V, Angi M, Romano MR, Filippelli ME, Di Iorio R, Costagliola C. Neurotrophic keratitis. Ophthalmologica. 2014;231(4):191–7. https://doi.org/10.1159/000354380. Epub 2013 Oct 2.
- Stevens S. Administering a subconjunctival injection. Community Eye Health. 2009;22(69):15. No abstract available.
- Baradaran-Rafii A, Eslani M, Haq Z, Shirzadeh E, Huvard MJ, Djalilian AR. Current and upcoming therapies for ocular surface chemical injuries. Ocul Surf. 2017;15(1):48–64. https://doi.org/10.1016/j.jtos.2016.09.002. Epub 2016 Sep 17.
- Eslani M, Baradaran-Rafii A, Movahedan A, Djalilian AR. The ocular surface chemical burns. J Ophthalmol. 2014;2014:196827. https://doi.org/10.1155/2014/196827. Epub 2014 Jul 1.

Recruiting

Stem Cell Therapy for Limbal Stem Cell Deficiency

ClinicalTrials.gov ID NCT03957954

Sponsor University of California, Los Angeles
Information provided by Sophie Deng, MD, PhD, University of California, Los Angeles (Responsible Party)
Last Update Posted 2023-01-11

Brief Summary
This phase I study will collect preliminary information on the activity and safety of cLSC. We will investigate the ability to manufacture and transplant cLSC onto the cornea successfully at the time of surgery (feasibility) and have cLSC begin to populate the ocular surface (efficacy) without serious adverse events (safety).

Detailed Description
20 patients with severe to total limbal stem cell deficiency (LSCD) in one eye attributable to injury or ocular surgery. The first 5 eyes will receive the cLSCs to

determine the feasibility and safety. Then the subsequent 15 eyes will be randomized into cLSC group and a scleral lens treatment (SCL) control group in a 2:1 ratio in blocks of 3 or 6 using a random number generator.

Official Title
Safety and Feasibility of Cultivated Autologous Limbal Stem Cells (cLSC) for Limbal Stem Cell Deficiency

Conditions
Limbal stem cell deficiency

Study Type
Interventional

Enrollment (Estimated)
20

Intervention/Treatment
Biological: cLSC
Device: Scleral contact lens (SCL)

Phase
Phase 1

Other Study ID Numbers
V001
R01EY021797 (U.S. NIH Grant/Contract)

Study Start (Actual)
2020-09-30

Primary Completion (Estimated)
2023-10-31

Study Completion (Estimated)
2023-12-31

Study Contact
Name: Clemence Bonnet, MD
Phone Number: 310-794-7813
Email: CBonnet@mednet.ucla.edu

Study Contact Backup
Name: Sofia Gudiel
Email: SGudiel@mednet.ucla.edu

United States

Eligibility Criteria
Ages Eligible for Study
18 years and older (adult, older adult)

Accepts Healthy Volunteer
No

Sexes Eligible for Study
All

Description

Main Inclusion Criteria
- Male or female ≥18 years of age.
- Best-corrected visual acuity in the affected eye of 20/200 or less.
- Documentation of a LSCD diagnosis and the central cornea is affected.
- Absence of lagophthalmos and eyelid abnormality.
- Adequate forniceal depth is ≥5 mm.
- LSCD fails to resolve by surgical treatments of the ocular surface during the previous 6 months of screening visit.
- If the etiology of LSCD is due to chemical injury, a minimal interval of 1 year since the initial chemical injury is required.
- A Schirmer test result at 5 min of ≥5 mm of wetting.
- Absence of active infectious keratitis in either eye at the enrollment visit.
- Have a life expectancy ≥2 years after enrollment.

Main Exclusion Criteria
- Pregnant, breastfeeding, or unwilling to practice birth control during participation in the study.
- Presence of a condition or abnormality that in the opinion of the Investigator would compromise the safety of the patient or the quality of the data.
- Exposure keratopathy or lagophthalmos of the study eye.
- Persistent severe ocular surface inflammation and/or meibomian gland dysfunction.
- Chemical injury occurred less than 12 months ago.
- Presence of ocular surface tumor.
- Uncontrolled diabetes with last hemoglobin A1C (HgA1C) >8.5.
- Presence of known allergies to any of the cLSC components.
- Current participation in another simultaneous medical investigation or trial.
- Unable to be compliant with or complete the requirements of the study.

Design Details
Primary Purpose: Treatment
Allocation: Randomized
Interventional Model: Parallel Assignment
Masking: None (open label)

Arms and interventions

Participant group/arm	Intervention/treatment
Experimental: cultivated limbal stem cells (cLSC) One dose of cultivated limbal stem cells (cLSC), size between 7.6 and 15 mm in the average diameter	Biological: cLSC • Transplantation of cLSC
Active comparator: scleral contact lens device (SCL) Scleral contact lens device (SCL) will be fitted to stabilize and improve ocular surface	Device: Scleral contact lens (SCL) • Scleral contact lens device (SCL) will be used to stabilize and improve ocular surface

Primary outcome measures

Outcome measure	Measure description	Time frame
Safety of cLSC transplant pertaining to infection and/or Corneal Perforation	Safety assessed by incidence of serious adverse events (SAEs) directly related to cLSC: infection and/or corneal perforation	12 months for all participants
Feasibility of cLSC transplant pertaining to manufacturing meeting release criteria	Feasibility assessed by successful manufacturing of cLSC from biopsy that meets release criteria	12 months for all participants
Feasibility of cLSC transplant pertaining to manufacturing without contamination	Feasibility assessed by successful cLSC manufacturing without contamination	12 months for all participants

Secondary outcome measures

Outcome measure	Measure description	Time frame
Changes in the area of corneal epithelial defect	Evaluate changes in the area of corneal epithelial defect (ED) or lack of ED	At 6 months and 12 months in both cLSC and the control groups
Changes in the clinical score	Evaluate changes in the clinical score determined by slit lamp examination using fluorescein staining and confocal imaging	At 6 months and 12 months in both cLSC and the control groups

Sponsor

University of California, Los Angeles

Collaborators

- National Eye Institute (NEI)
- California Institute for Regenerative Medicine (CIRM)

Investigators

- Principal Investigator: Sophie Deng, MD, PhD, Stein Eye Institute UCLA

General Publications

No publications available.

Canada

Recruiting

Autologous Cultured Corneal Epithelium (CECA) for the Treatment of Limbal Stem Cell Deficiency (CECA)

ClinicalTrials.gov ID NCT01756365

Sponsor CHU de Quebec-Universite Laval
Information provided by Ralph Kyrillos, CHU de Quebec-Universite Laval (Responsible Party)
Last Update Posted 2023-01-23

Brief Summary
The study "Autologous cultured corneal epithelium (CECA) for the treatment of corneal lesions associated with limbal stem cell deficiency" is the first clinical trial of this product manufactured at the LOEX laboratory. The culture of corneal epithelium strives to produce a reconstructed tissue with the therapeutical aim of treatment of limbal stem cell deficiency. The study is a phase I/phase II study with the goal to evaluate safety and efficacy of the CECA graft for the treatment of human patients suffering from limbal stem cell deficiency.

The trial is open to all genders. The inclusion of 5 minors is planned.

Official Title
Autologous Cultured Corneal Epithelium (Culture d'épithélium cornéen Autologue CECA) for the Treatment of Unilateral Corneal Lesions Associated With Limbal Stem Cell Deficiency

Conditions
Limbal Stem Cell Deficiency

Study Type
Interventional

Enrollment (Estimated)
54

Intervention/Treatment
Procedure: Surgical transplantation of ACCE

Phase
Not Applicable

Other Study ID Numbers
LOEX 015

Study Start
2012-12

Primary Completion (Estimated)
2025-12

Study Completion (Estimated)
2027-12

Canada
Québec, Canada, G1S 4L8
Recruiting
Centre universitaire d'Ophtalmologie CHU de Québec—HSS

Contact
Ralph Kyrillos, MD FRCS (C)
418-682-7573 ralph.kyrillos.1@ulaval.ca

Contact
Lucie Germain, PhD
4185254444 ext 61696 lucie.germain@fmed.ulaval.ca

Principal Investigator
Ralph Kyrillos, MD FRCS (C)

Ontario Locations

Toronto, Ontario, Canada, M5T 2S8
Recruiting
University Health Network–Toronto Western Hospital

Contact
Jennifer Coleman
4163404800 ext 5389

Principal Investigator
Allan Slomovic, MD, FRCSC

Quebec Locations

Montréal, Quebec, Canada, H1T 2M4
Recruiting
CIUSSS de l'Est de l'île de Montréal

Contact
Dorline Sokoundjou, MSc
514-252-3400 ext 7114 bfv2.hmr@ssss.gouv.qc.ca

Principal Investigator
Isabelle Brunette, MD

Eligibility Criteria
Ages Eligible for Study
1 year and older (child, adult, older adult)

Accepts Healthy Volunteer
No

Sexes Eligible for Study
All

Description

Inclusion Criteria
All Genders

- Adults
- Minors
- LSCD in one or two eyes. A minimum of 1–3 mm^2 of undamaged limbus is required for a biopsy to be taken without foreseeable adverse consequences for the donor eye

Exclusion Criteria
- Donor eye not sufficiently healthy to allow for the harvesting of a 1–3 mm^2 limbal biopsy without foreseeable consequences for the donor eye
- Pregnancy
- Breastfeeding
- Incapacitated person
- Known allergy to aprotinine (Trasylol®)
- Hypersensitivity to bovine proteins

Design Details
Primary purpose: treatment
Allocation: N/A
Interventional model: single-group assignment
Interventional Model Description: The culture of corneal epithelium strives to produce a reconstructed tissue with the therapeutical aim of treatment of limbal stem cell deficiency. The study is a phase I/phase II study with the goal to evaluate safety and efficacy of the CECA graft for the treatment of human patients suffering from limbal stem cell deficiency.
Masking: None (open label)

Arms and interventions

Participant group/arm	Intervention/treatment
Experimental: ACCE (autologous cultured corneal epithelium) graft for the treatment of corneal lesions Surgical transplantation of autologous cultured corneal epithelium	Procedure: Surgical transplantation of ACCE • The culture of corneal epithelium strives to produce a reconstructed tissue with the therapeutical aim of treatment of limbal stem cell deficiency

Primary outcome measures

Outcome measure	Measure description	Time frame
Anatomic assessment	Global severity score (epithelial deficit, corneal opacity, peripheral and central vascularization, integrity of the ocular surface)	1 year

Secondary outcome measures

Outcome measure	Measure description	Time frame
Best-corrected visual acuity	Evaluation according to Snellen chart	1 year
Level of pain	11-point numerical verbal scale	1 year
Quality-of-life measurements	Health utilities index mark 3	1 year

Sponsor
CHU de Quebec-Universite Laval

Collaborators
No information provided

Investigators
- Principal Investigator: Ralph Kyrillos, MD FRCS (C),CHU de Quebec

General Publications
No publications available.

China

Not Yet Recruiting

Safety and Efficacy of Pluripotent Stem Cell-derived Mesenchymal Stem Cell Exosome (PSC-MSC-Exo) Eye Drops Treatment for Dry Eye Diseases Post-Refractive Surgery and Associated With Blepharospasm

ClinicalTrials.gov ID NCT05738629

Sponsor Second Affiliated Hospital, School of Medicine, Zhejiang University
Information provided by Second Affiliated Hospital, School of Medicine, Zhejiang University (Responsible Party)
Last Update Posted 2023-02-22
Collaborators and Investigators

Brief Summary
The purpose of this study is to evaluate the preliminary safety and effectiveness of Pluripotent Stem Cell-derived Mesenchymal Stem Cell Exosome (PSC-MSC-Exo) Eye Drops in the treatment of dry eye diseases post refractive surgery and associated with blepharospasm.

Detailed Description
Dry eye disease (DED) is a multifactorial disease of the ocular surface characterized by instability of the tear film and the prevalence has been estimated to reach 20–40%. Especially, dry eye symptoms are common complaints of patients with blepharospasm or refractive surgery history, seriously affecting their quality of life. There

are many conventional therapies for DED which include ocular lubricants, oral essential fatty acid supplementation, lid hygiene and warm compresses, punctal occlusion, various treatments to obstructed meibomian glands, topical antibiotics, topical corticosteroids, topical secretagogues, topical non-glucocorticoid immunomodulatory drugs, and scleral contact lenses. However, the overall efficacy is still unsatisfactory with apparent side effects, long periods of treatment, and high costs. Mesenchymal stem cell-derived exosomes (MSC-Exo) are 30–150 nm microvesicles secreted by mesenchymal stem cells (MSCs), which can mediate the therapeutic efficacy of mesenchymal stem cells by the encapsulated proteins, miRNAs, and other bioactive substances, and showed potential in the treatment of various diseases. This study aims to evaluate the preliminary safety and effectiveness of PSC-MSC-Exo in the treatment of dry eye diseases post-refractive surgery and associated with blepharospasm.

This is an open-label, single-arm, before-after study with 12 subjects with dry eye diseases and the history of refractive surgery or blepharospasm. There are 5 visits over the course of 12 weeks. Visit 1: Week-1 Visit 2: Week-2 Visit 3: Week-4 Visit 4: Week-8 Visit 5: Week-12

Official Title

A Clinical Study of the Safety and Efficacy of Pluripotent Stem Cell-derived Mesenchymal Stem Cell Exosome (PSC-MSC-Exo) Eye Drops Treatment for Dry Eye Diseases Post Refractive Surgery and Associated With Blepharospasm Dry Eye Disease

Study Type

Interventional

Enrollment (Estimated)

12

Intervention/Treatment

Drug: Pluripotent Stem Cell-derived Mesenchymal Stem Cell Exosome (PSC-MSC-Exo) Eye Drops

Phase

Phase 1 Phase 2

Other Study ID Numbers

2022-0290

Study Start (Estimated)

2023-03

Primary Completion (Estimated)

2025-02

Study Completion (Estimated)

2025-02

Study Contact
Name: Qi Gao, Doc.
Phone Number: 86-15088774002
Email: gaoqi8977@zju.edu.cn

China
Hanzhou, Zhejiang, China, 310000
2nd Affiliated Hospital, School of Medicine, Zhejiang University

Contact
Qi Gao, Doc
86-15088774002

Eligibility Criteria
Ages Eligible for Study
18–70 years (adult, older adult)

Accepts Healthy Volunteer
No

Sexes Eligible for Study
All

Description

Inclusion Criteria
- Clinical diagnosis of apparent dry eye symptoms by ophthalmologists.
- Patients experience dry eye symptoms with a definitive history of FS-LASIK or SMILE surgery, or blepharospasm, and previous application of artificial tears for more than 3 months could not relieve the dry eye symptoms.
- Have the following symptoms in at least one eye: dryness, burning sensation, foreign body sensation, discomfort in the ocular surface, or visual fatigue.
- Tear secretion test (Schirme's test) of either eye ≤10 mm/5 min.
- The corneal fluorescein staining score (FLCs) was more than 3 and less than 6.
- Tear break-up time (TBUT) is less than 10 s.

Exclusion Criteria
- Those who are allergic to any component of the drug in this study.
- Pregnant or nursing women.
- Patients with active fungal, bacterial, or viral keratitis or conjunctivitis.
- Have serious heart, lung, liver, or kidney diseases.
- Other incurable ocular diseases before the study, such as glaucoma, uveitis, retinitis pigmentosa.
- Wearing contact lenses and unwilling to take off in the study.
- Performed ocular surgery (including cataract surgery) in recent 3 months.
- Enrolled in other interventional clinical studies at the same time.
- Application of eye drops that might affect the clinical study in the past 24 h.
- Unable to complete the study according to the investigators' requirements.
- Application of eye drops other than artificial tears.

- Application of systemic medication known to reduce tear production, such as anti-anxiety drugs, antipsychotics and steroids.
- Eyelids or eyelashes anomaly.
- Serious systemic diseases.
- Refused to sign the informed consent form to participate in the experiment.

Design Details

Primary purpose: treatment
Allocation: N/A
Interventional model: single-group assignment
Masking: none (open label)

Arms and interventions

Participant group/arm	Intervention/treatment
Experimental: PSC-MSC-Exo eye drops treatment Participants will receive artificial tears for 2 weeks to get the normalized baseline, followed by PSC-MSC-Exo eye drop intervention for 12 weeks	Drug: pluripotent stem cell-derived mesenchymal stem cell exosome (PSC-MSC-Exo) eye drops • Participants will receive artificial tears for 2 weeks to get the normalized baseline, followed by PSC-MSC-Exo eye drops 0.125 mL/single eye/one time, four times a day for 12 weeks. The follow-up visit will be 12 weeks since the administration of PSC-MSC-Exo eye drops

Primary outcome measures

Outcome measure	Measure description	Time frame
Change in ocular surface disease index (OSDI)	The OSDI questionnaire contains 12 questions to measure dry eye symptoms. OSDI scores range from 0 to 100. The lower the score, the more alleviation of dry eye symptoms. Compared with the baseline, a reduction of more than 10 points indicates improvement	1 week, 2 weeks, 4 weeks, 8 weeks, 12 weeks

Secondary outcome measures

Outcome measure	Measure description	Time frame
Tear secretion test (Schirmer test)	A paper strip is placed on the eye under the lower lid for a specified time period. The length of the strip wetted by the tears was measured in millimeters. A positive change from baseline indicates an improvement in each eye	1 weeks, 2 weeks, 4 weeks, 8 weeks, 12 weeks
Tear film break-up time	Time required for dry spots to appear on the surface of the eye after blinking was measured in seconds. A positive change from baseline indicates improvement in each eye	1 weeks, 2 weeks, 4 weeks, 8 weeks, 12 weeks

Ocular surface staining score	Damage of the ocular surface was measured by slit-lamp examination after staining with non-toxic fluorescein sodium. Corneal and conjunctival staining were graded as per the National Eye Institute (NEI) pictorial scale. The lower the score, the less signs of dry eye disease a patient exhibits. A negative change from baseline indicates an improvement in each eye	1 weeks, 2 weeks, 4 weeks, 8 weeks, 12 weeks
Best-corrected visual acuity (BCVA)	Whether exosomes affect visual acuity	1 weeks, 2 weeks, 4 weeks, 8 weeks, 12 weeks
Conjunctiva redness score	Conjunctival hyperemia was graded in each eye under slit-lamp magnification by assigning a score of 0–4 for each quadrant based on comparison to the Allergan Dry Eye Redness Scale with photographic reference where: 0 = normal, vessels of bulbar conjunctiva are easily observed; 1 = trace redness; 2 = mild redness; 3 = moderate redness; 4 = severe redness, and each score is associated with a reference photo. A negative change from baseline indicates an improvement in each eye	1 weeks, 2 weeks, 4 weeks, 8 weeks, 12 weeks
Tear meniscus height	The distance between the line of reflection along the top of the tear prism to the edge of the eyelid was measured in millimeters. A positive number change from baseline indicates improvement	1 weeks, 2 weeks, 4 weeks, 8 weeks, 12 weeks
Corneal confocal microscopy	The morphology, distribution, and quantity of corneal nerve fibers were recorded by confocal image after LASIK and SMILE	1 weeks, 2 weeks, 4 weeks, 8 weeks, 12 weeks
Meibomian gland expressibility scores	Squeeze the central meibomian glands of the lower eyelid and assess the expressibility of the five central glands, where 0 = all glands have meibum extruded, 1 = 3–4 glands have meibum extruded, 2 = 1–2 glands have meibum extruded, 3 = no glands have meibum extruded	1 weeks, 2 weeks, 4 weeks, 8 weeks, 12 weeks
Meibomian glands photography	The morphology, quantity, and blockage of meibomian gland can be assigned a score of 0–3 points according to meibomian gland dropout, where 0 = normal without dropout; 1 = meibomian gland dropout<1/3; 2 = meibomian gland dropout 1/3–2/3; 3 = meibomian gland dropout >2/3. The total score of upper and lower meibomian glands is 6 points, of which Grade 0 = 0–1 point; Grade 1 = 2–3 points; Grade 2 = 4–5 points; Grade 3 = 6 points	1 weeks, 2 weeks, 4 weeks, 8 weeks, 12 weeks

Sponsor
Second Affiliated Hospital, School of Medicine, Zhejiang University

Collaborators
- Zhejiang University
- Hangzhou Yuansheng biotechnology Co., Ltd

Investigators
- Study Chair: Juan Ye, Doc., 2nd Affiliated Hospital, School of Medicine, Zhejiang University

General Publications
No publications available.

Recruiting

Effect of UMSCs Derived Exosomes on Dry Eye in Patients with cGVHD

ClinicalTrials.gov ID NCT04213248

Sponsor Zhongshan Ophthalmic Center, Sun Yat-sen University
Information provided by Zhongshan Ophthalmic Center, Sun Yat-sen University (Responsible Party)
Last Update Posted 2022-02-11

Brief Summary
The purpose of this study is to determine whether umbilical mesenchymal stem cells (UMSCs) derived exosomes could alleviate dry eye symptoms in patients with chronic graft-versus-host diseases (cGVHD).

Detailed Description
Among patients with cGVHD, 60–90% are affected by dry eye symptoms, presenting as progressive development of dryness, foreign body sensation, photophobia, pain, and even blindness, which seriously affect their life quality. At present, the standard first-line treatment of cGVHD is still hormone or combined with immunosuppressant of cyclosporine. The dry eyes related to cGVHD are treated with artificial tears, lacrimal punctum embolization, local immunosuppressant, or even blepharoplasty to reduce dry eye symptoms. Although a variety of new immunosuppressants and monoclonal antibodies have been used in clinic, the overall efficacy is still unsatisfactory with apparent side effects, and the period of treatment is long with high costs. Therefore, it is an urgent task for clinicians to explore new methods of cGVHD-related dry eye therapy and improve the survival rate and quality of life of patients.

Exosomes are a kind of membrane vesicles secreted by parental cells, which can mediate the transfer of RNA, protein, DNA, and other functional molecules between cells and regulate the function of target cells. The therapeutic potential of UMSC exosomes has been widely studied in diseases of liver, kidney, skin, and so on. Previously, we found that UMSC exosomes can significantly alleviate the symptoms of dry eye in the dry eye animal model. At present, we plan to clarify its efficacy in clinical research.

The main objective of this study is to assess the alleviation of dry eye symptoms in patients with cGVHD after UMSC-exo treatment by measuring Ocular Surface Index Score (OSDI), and the second objective includes the measurement of tear secretion amount, tear break time, the areas stained by fluorescent, ocular redness, tear meniscus, and best-corrected visual acuity.

Approximately 27 study subjects will be recruited. The treatment group will receive artificial tears for 2 weeks to normalize the baseline, followed by intervention of UMSC-exo 10 ug/drop, four times a day for 14 days. The follow-up visit will be 12-week; progression of dry eye will be measured.

Official Title

Effect of Umbilical Mesenchymal Stem Cells Derived Exosomes on Dry Eye in Patients With Chronic Graft Versus Host Diseases Dry Eye

Study Type

Interventional

Enrollment (estimated)

27

Intervention/Treatment

Drug: Umbilical mesenchymal stem cell-derived exosomes

Phase

Phase 1 Phase 2

Other Study ID Numbers

2019KYPJ048

Study Start (Actual)

2020-02-21

Primary Completion (Estimated)

2023-05-18

Study Completion (Estimated)

2023-12

Study Contact

Name: Tian Zhou
Phone Number: (020)66615460
Email: zhoutian@gzzoc.com
China

Guangzhou, Guangdong, China, 510000
Recruiting
Zhongshan Ophthalmic Center

Contact
Tian Zhou, phD
zhoutian@gzzoc.com

Eligibility Criteria
Ages Eligible for Study
18–70 years (adult, older adult)

Accepts Healthy Volunteer
No

Sexes Eligible for Study
All

Description

Inclusion Criteria
Diagnosed cGVHD with apparent dry eye symptoms by intern specialists and ophthalmologists, and previous application of artificial tears could not relieve the dry eye symptoms.
And meet the following standards:

- Have the following symptoms in at least one eye: dryness, burning sensation, foreign body sensation, discomfort in the ocular surface, or visual fatigue.
- Tear secretion test (Schirmer's test) of either eye ≤10 mm/5 min.
- The corneal fluorescein staining score (CFS) was more than 4.
- Tear break time (TBUT) is less than 10 s.

Exclusion Criteria
- Those who are allergic to any component of the drug in this study.
- Pregnant or nursing women.
- Patients with active fungal, bacterial, or viral keratitis or conjunctivitis.
- Have serious heart, lung, liver, or kidney diseases.
- Other incurable ocular diseases before the study, such as glaucoma, uveitis, retinitis pigmentosa.
- Wearing contact lenses and unwilling to take off in the study.
- Performed ocular surgery (including cataract surgery) in recent 3 months.
- Enrolled in other interventional clinical studies at the same time.
- Application of eye drops that might affect the clinical study in the past 24 h.
- Unable to complete the study according to the investigators' requirements.
- Serious systemic diseases.

Design Details
Primary purpose: treatment
Allocation: N/A
Interventional model: single-group assignment
Masking: none (open label)

Arms and interventions

Participant group/arm	Intervention/treatment
Experimental: UMSC-exo treatment Participants will receive artificial tears for 2 weeks to get the normalized baseline, followed by UMSC-exo intervention for 2 weeks	Drug: umbilical mesenchymal stem cells derived exosomes • Participants will receive artificial tears for 2 weeks to get the normalized baseline, followed by UMSC-exo 10 ug/drop, four times a day for 14 days. The follow-up visit will be 12 weeks • Other Names: – UMSC-exo

Primary outcome measures

Outcome measure	Measure description	Time frame
Changes in ocular surface disease index (OSDI) score	The OSDI is a 12-question validated questionnaire used to measure dry eye symptoms. The OSDI scoring scale ranges from 0 to 100. The lower the score, the more symptomatic relief from dry eye symptoms a patient experiences. Baseline-adjusted scores were tabulated; a decrease over 10 scores from baseline indicates an improvement	3 days, 2 weeks, 3 weeks, 4 weeks, 6 weeks, 8 weeks, 10 weeks, 12 weeks

Secondary outcome measures

Outcome measure	Measure description	Time frame
Changes in tear secretion amount by Schirmer's test	The investigator placed a paper strip on the eye under the lower lid for a specified time period. The length of the strip wetted by the tears was measured in millimeters. Baseline-adjusted values were tabulated; a positive number change from baseline indicates an improvement in each eye individually	3 days, 2 weeks, 3 weeks, 4 weeks, 6 weeks, 8 weeks, 10 weeks, 12 weeks
Changes in tear break time	Time required for dry spots to appear on the surface of the eye after blinking was measured in seconds. Baseline-adjusted scores were tabulated; a positive number change from baseline indicates improvement in each eye individually	3 days, 2 weeks, 3 weeks, 4 weeks, 6 weeks, 8 weeks, 10 weeks, 12 weeks

Changes in ocular surface staining	Damage to the ocular surface was assessed using non-toxic ophthalmic dye during slit-lamp review. Corneal and conjunctival staining were graded as per the National Eye Institute (NEI) pictorial scale. The lower the score, the less signs of dry eye disease a patient exhibits. Baseline-adjusted scores were tabulated; a negative number change from baseline indicates an improvement in Each eye individually Damage to the ocular surface was assessed using non-toxic ophthalmic dye during slit-lamp review. Corneal and conjunctival staining were graded as per the National Eye Institute (NEI) pictorial scale. The lower the score, the less signs of dry eye disease a patient exhibits. Baseline-adjusted scores were tabulated; a negative number change from baseline indicates an improvement in each eye individually	3 days, 2 weeks, 3 weeks, 4 weeks, 6 weeks, 8 weeks, 10 weeks, 12 weeks
Changes in best-corrected visual acuity (BCVA)	To understand the effect of exosomes on visual acuity	3 days, 2 weeks, 3 weeks, 4 weeks, 6 weeks, 8 weeks, 10 weeks, 12 weeks
Changes in conjunctiva redness score	Description: To explore the effect of exosomes on conjunctiva. Nasal and temporal bulbar conjunctival hyperemia were graded separately in each eye by the investigator under slit-lamp magnification by assigning a score of 0 to 4 for each quadrant based on comparison to the allergan dry eye redness scale with photographic reference where: 0 = normal, vessels of bulbar conjunctiva are easily observed; 1 = trace redness; 2 = mild redness; 3 = moderate redness; 4 = severe redness, and each score is associated with a reference photo	3 days, 2 weeks, 3 weeks, 4 weeks, 6 weeks, 8 weeks, 10 weeks, 12 weeks

Changes in tear meniscus height	Description: to reflect the effect of exosomes on tear production. The distance between the line of reflection along the top of the tear prism to the edge of the eyelid was measured in millimeters. Baseline-adjusted scores were tabulated; a positive number change from baseline indicates improvement. Each eye was assessed individually. Both eyes contributed to the mean	3 days, 2 weeks, 3 weeks, 4 weeks, 6 weeks, 8 weeks, 10 weeks, 12 weeks

Sponsor
Zhongshan Ophthalmic Center, Sun Yat-sen University

Collaborators
No information provided

Investigators

- Principal Investigator: Xialin Liu, Prof. Zhongshan Ophthalmic Center, Sun Yat-sen University

General Publications
No publications available.

Not Yet Recruiting

Therapeutic Effect of Stem Cell Eye Drops on Dry Eye Disease

ClinicalTrials.gov ID NCT05784519

Sponsor The Affiliated Nanjing Drum Tower Hospital of Nanjing University Medical School
Information provided by The Affiliated Nanjing Drum Tower Hospital of Nanjing University Medical School (Responsible Party)
Last Update Posted 2023-03-27

Brief Summary
The goal of this non-randomized, prospective, open, one-arm clinical study is to learn about the clinical efficacy of stem cell eye drops in patients with dry eye disease (DED) who failed to respond to artificial tear sodium hyaluronate eye drops three times a day for 2 weeks.

The main question aims to answer are:

- How effective are stem cell eye drops in patients with DED?
- How safe are stem cell eye drops for patients with DED? Participants will be treated with mesenchymal stem cells (MSCs) eye drops, $5 \times 10^5/50$ μl in each eye, twice a day for 2 weeks and they will be followed up for 3 months after treatment.

Detailed Description

This study was mainly divided into three parts: screening period, treatment period, and late treatment period. (1) Screening period: For the DED patients with artificial tear sodium hyaluronate drops three times a day and no effect for two consecutive weeks, OSDI score, fluorescein sodium stain score, tear secretion test, and tear film rupture time examination were performed. Ten eligible subjects were screened out according to the inclusion/exclusion criteria, and the subjects were asked to sign informed consent. Relevant clinical examination data were collected and subjects were enrolled. (1) OSDI questionnaire: It was divided into 3 dimensions, including ocular symptoms, visual function, and environmental factors. (2) Fluorescein staining score of ocular surface: After staining with 0.25% fluorescein sodium, defect staining was observed under slit-lamp microscope. Cornea was scored in 3 areas: top, center and bottom. Each area was scored by 0 points (without any staining), 1 point (partial staining), 2 points (staining of more than half of the area), and 3 points (staining of the whole area). The total was divided into the sum of 3 areas. A higher score indicates more severe ocular surface damage. (3) Schirmer test: Standard tear test paper was taken and held in the conjunctival sac at one-third of the lower outer eyelid, and the patient was instructed to close the eyes gently. The results were observed 5 min later. The tear secretion test <10 mm/5 min was abnormal. (4) Tear break-up time: After staining with 0.25% fluorescein sodium, patients were instructed to close their eyes under the slit lamp microscope. The tear film rupture time was calculated from the time of eye opening to the appearance of the first burst spot, and the average value was taken for three consecutive tests. The rupture time of tear film was less than 10 s. (2) Treatment period: Formal clinical treatment: 10 enrolled patients received MSC eye drops twice a day for 2 weeks. OSDI score, ocular fluorescein sodium stain score, tear secretion test, and tear film rupture time examination were performed every weekend during drug administration. Adverse events, serious adverse events, and drug-related adverse events ≥III during treatment were recorded to evaluate drug safety. (3) Late treatment: 3 months of follow-up after the end of treatment. OSDI score was performed during weekly telephone follow-up, and clinical indicators (OSDI score, fluorescein sodium staining score of ocular surface, tear secretion test, and tear film rupture time test) were examined every 4 weeks to evaluate the treatment effect of DED.

Official Title

Clinical Study of Mesenchymal Stem Cell Eye Drops for Treating Dry Eye Disease

Conditions

Mesenchymal stem cell
Dry eye syndromes

Study Type

Interventional

Enrollment (Estimated)

10

Intervention/Treatment

Drug: MSCs eye drops

Phase
Early phase 1

Other Study ID Numbers
SC202200102

Study Start (Estimated)
2023-06-01

Primary Completion (Estimated)
2024-07-31

Study Completion (Estimated)
2025-07-31

Study Contact
Name: Kai Hu, doctor
Phone Number: +86 13951606283
Email: kai_hu@nju.edu.cn

Eligibility Criteria
Ages Eligible for Study
40–60 years (adult)

Accepts Healthy Volunteer
No

Sexes Eligible for Study
Female

Description

Inclusion Criteria
- As determined by the investigator, the subject is able to understand and comply with the protocol requirements.
- Subject or subject's legal representative (if applicable) sign and date the written informed consent or any privacy authorization document required prior to the commencement of the study process.
- Age 40–60.
- Female patients.
- The subject has received artificial tear sodium hyaluronate treatment with eye drops three times a day, but the treatment has no effect for 2 weeks, and the clinical examination results are consistent with:
  - OSDI score ≥13.
  - Schirmer test <10 mm/5 min.
  - Tear break-up time <10 s.
  - Fluorescein sodium staining on the ocular surface was positive.

Exclusion Criteria

- Eye exclusion criteria:
  - Subjects with other eye diseases such as glaucoma, cataract, uveitis, and optic neuritis.
  - The subjects had undergone eye surgery (e.g., cataract surgery) within the last 3 months.
  - Subjects who wear contact lenses for a long time and are unwilling to remove them in the study.
  - The subject has received eye drops in the past 24 h that may affect the clinical study.
- Exclusion criteria related to infectious diseases:
  - The subject has fungal, bacterial, or viral keratitis or conjunctivitis with evidence of infection.
  - The subject has chronic hepatitis B virus (HBV) infection or chronic hepatitis C virus (HCV) infection.
  - The subject has any identified congenital or acquired immunodeficiency (e.g., common immunodeficiency, human immunodeficiency HIV infection, organ transplantation).
  - The subject has active tuberculosis (TB positive).
  - Subjects receive any live vaccine within 30 days prior to screening.
- Exclusion criteria for general circumstances:
  - The subject is allergic to the stem cell eye drops.
  - The subject has any unstable or uncontrolled cardiovascular, pulmonary, liver, kidney, gastrointestinal, urogenital, hematological, clotting, immunological, endocrine/metabolic, or other medical condition that the investigator deems to interfere with the study or endanger the safety of the subject.
  - Subjects had had any surgery requiring general anesthesia within 30 days prior to enrollment, or planned to have a larger procedure during the study period surgery.
  - Subjects have a history of severe neurological diseases, including stroke, multiple sclerosis, brain tumors, or neurodegenerative diseases.
  - The subjects had active psychosis, which the researchers thought might interfere with their compliance with the study process.
  - Lactating female subjects, or female subjects with positive serum pregnancy test results during the screening period, or positive urine pregnancy test results prior to study drug administration.

Design Details

Primary purpose: treatment
Allocation: N/A
Interventional model: single-group assignment
Masking: none (open label)

Arms and interventions

Participant group/arm	Intervention/treatment
Experimental: experimental group The MSCs eye drops was administered to enrolled patients with 5 × 10^5 /50 μl in each eye, twice a day for 2 weeks	Drug: MSCs eye drops • MSCs were used to prepare single-cell suspension, and 0.9% sodium chloride was added to balance PH value to prepare the effective, safe, and stable eye drops. Giving the patient with 5 × 10^5 /50 μl in each eye, twice a day for 2 weeks

Primary outcome measures

Outcome measure	Measure description	Time frame
The change of OSDI score	OSDI questionnaire: It was divided into 3 dimensions, including ocular symptoms, visual function, and environmental factors. The total score is less than 13	1 week
The change of Schirmer test	Standard tear test paper was taken and held in the conjunctival sac at 1/3 of the lower eyelid, and the patient was instructed to close the eyes gently. The results were observed 5 min later. Schirmer test <10 mm/5 min was abnormal	1 week

Secondary outcome measures

Outcome measure	Measure description	Time frame
Tear break-up time	After staining with 0.25% fluorescein sodium, patients were instructed to close their eyes under the slit lamp microscope. The tear film rupture time was calculated from the time of eye opening to the appearance of the first burst spot, and the mean value was taken for three consecutive tests. The tear film rupture time <10 s is abnormal	1 week
Fluorescein sodium staining of the ocular surface	After staining with 0.25% fluorescein sodium, defect staining was observed under slit-lamp microscope. Cornea was scored in 3 areas: top, center, and bottom. Each area was scored according to 0 points (without any staining), 1 point (partial staining), 2 points (staining of more than half of the area), and 3 points (staining of the whole area), and the total was divided into the sum of 3 areas. A higher score indicates more severe ocular surface damage	1 week

Collaborators and Investigators

This is where you will find people and organizations involved with this study.

Sponsor

The Affiliated Nanjing Drum Tower Hospital of Nanjing University Medical School

Collaborators

No information provided

Investigators
No information provided

General Publications
No publications available.

France

Recruiting

Cultured Autologous Oral Mucosa Epithelial Sheet for the Treatment of Bilateral Limbal Stem Cell Deficiency (FEMJA)

ClinicalTrials.gov ID NCT03949881

Sponsor Hospices Civils de Lyon
Information provided by Hospices Civils de Lyon (Responsible Party)
Last Update Posted 2023-04-24

Brief Summary
Some severe ocular burns or other rare ocular pathologies can be associated with a total loss of corneal epithelial stem cells (i.e., limbal stem cell deficiency—LSCD), which leads to cornea invasion by the conjunctiva and a subsequent opacification. When LSCD is total and bilateral, both eyes are affected leading to full blindness and a poor quality of life, with a paradoxical photophobia that may be painful. Fewer than 100 patients bear this rare condition in France.

When patients suffer from total and bilateral LSCD, no treatment has been proven to provide clinical benefits: contralateral limbus is unavailable for autologous limbus graft or autologous limbal stem cells culture; allogeneic limbus graft requires immunosuppressive treatment leading to too important serious adverse effects compared to the expected benefit, and does not last long (<2 years); and allogenic cornea transplantation is impossible since always rejected due to neovascularization.

A new way to treat these patients is to cultivate autologous corneal-like epithelium, and to graft it in order to restore transparency and to allow, if needed, a complementary corneal graft. Such an epithelium can be produced from autologous jugal mucosa cells. Epithelial jugal mucosa sheets transplantation has been assessed in a phase I/II clinical trial on 26 patients which showed that it is well-tolerated and effective but the culture technology used in this clinical trial is no longer available. A new enzymatic detachment process has been developed by the Hospices Civils de Lyon. Proof of concept was obtained from both in vitro and ex vivo studies: detachment with collagenase at 0.5 mg/mL does not damage basement membrane proteins, so collagenase 0.5 mg/mL-detached FEMJA were found to adhere, continue to ensure renewal of the differentiated epithelium 15 days after grafting onto an ex vivo porcine de-epithelialized stroma model.

Considering these results, we aim to perform a clinical trial in order to evaluate tolerance and efficacy of the autologous jugal mucosa cell sheet (Feuillets Epithéliaux de Muqueuse Jugale Autologue—FEMJA) cultured with this innovative process.

Official Title
Cultured Autologous Oral Mucosa Epithelial Sheet for the Treatment of Bilateral Limbal Stem Cell Deficiency FEMJA for "Feuillet Epithélial de Muqueuse Jugale Autologue"

Conditions
Total bilateral limbal cell deficiency

Study Type
Interventional

Enrollment (Estimated)
40

Intervention/Treatment
Biological: FEMJA transplantation

Phase
Phase 1 Phase 2

Other Study ID Numbers
69HCL19_0032

Study Start (Actual)
2020-11-13

Primary Completion (Estimated)
2028-05-13

Study Completion (Estimated)
2028-05-13

Study Contact
Name: Carole Burillon, MD
Phone Number: 472.11.62.17 ext +33
Email: carole.burillon@chu-lyon.fr

Study Contact Backup
Name: Celine Auxenfans
Phone Number: 472 11 04 65 ext +33
Email: celine.auxenfans@chu-lyon.fr
France

Lyon, France, 69003
Recruiting
Ophthalmology Department, Edouard Herriot hospital, Hospices Civils de Lyon

Contact
Carole Burillon, MD
472.11.62.17 ext +33 carole.burillon@chu-lyon.fr

Principal Investigator
Carole Burillon, MD

Eligibility Criteria
Ages Eligible for Study
18 years and older (adult, older adult)

Accepts Healthy Volunteer
No

Sexes Eligible for Study
All

Description

Inclusion Criteria
- Male or female aged ≥18 years
- Signed and dated informed consent for participation in the study
- Total bilateral limbal stem cell deficiency
- Caused by thermal or chemical burn, cornea transplantation, and other bilateral disorders of the ocular surface
- Severe loss of vision (<2/10 on decimal scale or/and EDTRS)
- The subject must be covered by a social security system

Exclusion Criteria
- Eye inflammation
- Strictly unilateral ocular affliction
- Acute systemic infection, objectified during consultation by the investigator and on the following paraclinical parameter: C Reactive Protein rate
- History of acute phase of ocular inflammation in the previous year
- History of neoplastic disease
- Glaucoma defined as intraocular pressure (air tonometer and applanation tonometer) ≥22 mmHg
- Total symblepharon (comprising eyelid aperture): impossibility to open the 2 eyes
- History of hypersensibility or allergy to antibiotics or serum
- Women who are or may be pregnant or breastfeeding
- Patients with any active infectious disease (HBV, HCV, HIV, HTLV-1, and syphilis)
- Patients who are otherwise ineligible for participation in the study in the opinion of the investigator
- Delay of less than 1 year after chemical or thermal burns
- Person under judicial protection
- Contraindication related to anesthesia
- Contraindication to fluoresceine

- Oral mucosa tumor, pharynx, or larynx tumor
- Fungal or viral infection of the ENT area
- Bacterial infection of the oral mucosa, pharynx, or larynx
- Hypersensitivity to ofloxacin (or other quinolone drugs), fluorometholone, or betamethasone
- Hypersensitivity to one of the excipients of the eye drops used: Monosodium phosphate, anhydrous disodium phosphate, polysorbate 80, sodium chloride, sodium edetate, polyvinyl alcohol, benzalkonium chloride, hydroxypropylmethylcellulose, hydrochloric acid, sodium hydroxide, lactose, cellulose, crospovidone, aspartam, magnesium stearate
- Phenylketonuria
- Psychotic states not yet controlled by treatment;
- Vaccination with a live vaccine
- Bacterial, fungal, myco-bacterial infection of eye structures
- Some evolving viral infections (including hepatitis, herpes, chickenpox, shingles) or other viral infections of the cornea or conjunctiva (except for a keratitis to Herpes zoster)
- Patients treated with Class IA or III antiarrhythmics, tricyclic antidepressants, macrolides, and antipsychotics

Design Details

Primary purpose: treatment
Allocation: N/A
Interventional model: single-group assignment
Masking: none (open label)

Arms and interventions

Participant group/arm	Intervention/treatment
Experimental: FEMJA transplantation FEMJA epithelium is obtained by culturing oral mucosal epithelial cells without any need of support substrates, or carriers, and the transparent fabricated sheets show strong, rapid adhesion on corneal stroma in vivo, without any need for suturing The cultivated oral mucosal epithelium will be directly grafted onto the corneal stroma. The sheet is grafted without suture onto the exposed stromal bed after removal of the conjunctiva and fibrosis from the cornea. The grafted corneal surface is then covered with a soft permanent contact lens for protection during healing (between 3–15 days, according to tolerance). If the stroma appears opaque because of deep stromal scars a corneal allograft will be performed 12 months after FEMJA transplantation	Biological: FEMJA transplantation • The cultivated oral mucosal epithelium will be directly grafted onto the corneal stroma. The sheet is grafted without suture onto the exposed stromal bed after removal of the conjunctiva and fibrosis from the cornea. The grafted corneal surface is then covered with a soft permanent contact lens for protection during healing (between 3–15 days, according to tolerance). If the stroma appears opaque because of deep stromal scars a corneal allograft will be performed 12 months after FEMJA transplantation • Other Names: – Cultivated oral mucosal epithelium transplantation

Primary outcome measures

Outcome measure	Measure description	Time frame
Number of patients with any improvement in visual acuity	Visual acuity measured at baseline (preoperative status) and 24 months after a jugal mucosa epithelial cell sheet transplantation (FEMJA transplantation), according to the Early Treatment Diabetic Retinopathy Study (ETDRS) chart. ETDRS LogMAR tests allow measurement of very low visual acuity	24 months after FEMJA transplantation

Secondary outcome measures

Outcome measure	Measure description	Time frame
Number of successful FEMJA transplantations	A success is defined by obtaining at least one FEMJA sheet that could be grafted without perforating it	During 1 day of transplantation
Number of adverse events related to the procedure	Adverse events related to the procedure; especially expected adverse events: inflammation, infection, and perforation	Up to 24 months after FEMJA transplantation
Improvement in physical signs	Any improvement from inclusion in physical signs measured with a slit lamp, in terms of: persistent epithelial defect; superficial punctate epithelial keratitis; and conjunctival epithelium on cornea	6 months, 12 months, 18 months, and 24 months after FEMJA transplantation
Improvement in neovascularization	Number of patients with a decrease in the number of vascular pediculus near the limbus and their activity, compared to inclusion	6 months, 12 months, 18 months and 24 months after FEMJA transplantation
Improvement in symptoms, i.e., functional signs	Number of patients with an improvement in functional signs measured with Likert scales by self-reporting from the patient, in terms of photophobia, eye dryness, and pain, compared to inclusion	6 months, 12 months, 18 months and 24 months after FEMJA transplantation
Corneal graft rejection	Among patients with secondary corneal graft, number of patients with graft rejection defined by intraocular pression ($\geq$22 mmHg) and significant increase in the thickness of the stroma of the cornea compared to the measured value just post graft on corneal topography	24 months after FEMJA transplantation
Correlation between biological FEMJA characteristics and changes in visual acuity	FEMJA characteristics (extraction yield, percentage of proliferative epithelial cells) will be compared according to the clinical response for changes in visual acuity	24 months after FEMJA transplantation

Sponsor
Hospices Civils de Lyon

Collaborators
No information provided

Investigators
No information provided

General Publications
No publications available.

Spain

Not Yet Recruiting

Cultivated Limbal Epithelial Transplantation (CLET) for Limbal Stem Cell Deficiency (LSCD) (CLET-4-LSCD)

ClinicalTrials.gov ID NCT03884569

Sponsor Instituto Universitario de Oftalmobiología Aplicada (Institute of Applied Ophthalmobiology)—IOBA
Information provided by Instituto Universitario de Oftalmobiología Aplicada (Institute of Applied Ophthalmobiology)—IOBA (Responsible Party)
Last Update Posted 2022-02-07

Brief Summary
CLET is a published treatment for the management of corneal failure due to extensive LSCD. Due to our previous studies on this novel treatment, the regulatory agency of Spain "Agencia Española de Medicamentos y Productos Sanitarios (AEMPS)" authorized our institution (IOBA-University of Valladolid) to perform this kind of therapy (CLET) in a case-by-case base following the Special Situation Medicines Policy Procedure in Spain. Upon approval of the permanent authorization patients will be included as specified by AEMPS. The objective of this study is to perform a protocolized treatment and follow-up so that results can be reported to the scientific community.

Official Title
Cultivated Limbal Epithelial Transplantation (CLET) for Limbal Stem Cell Deficiency (LSCD): an Observational Study on Safety and Effectiveness

Conditions
Limbal stem cell deficiency

Study Type
Observational

Enrollment (Estimated)
30

Intervention/Treatment
Other study ID numbers
IOBA201901

Study Start (Estimated)
2022-04-01

Primary Completion (Estimated)
2022-12-31

Study Completion (Estimated)
2022-12-31

Study Contact
Name: Margarita Calonge, MD, PhD
Phone Number: 34983184750
Email: calonge@ioba.med.uva.es
Spain

Valladolid, Spain, 47011
IOBA

Contact
Margarita Calonge, MD, PhD

Principal Investigator
Margarita Calonge, MD, PhD

Eligibility Criteria
Ages Eligible for Study
(child, adult, older adult)

Accepts Healthy Volunteer
No

Sexes Eligible for Study
All

Sampling Method
Non-probability sample

Study Population
Patients diagnosed with corneal failure due to LSCD who accept and sign the informed written consent treatment to be treated with CLET

Description
Total and/or severe LSCD of any etiology in one or both eyes.

Design Details
Observational model: cohort

Time perspective: other

Cohorts and interventions

Group/cohort
Patients treated with CLET Previously treated with CLET patients are included retrospectively, as the follow-up procedure is already stablished in our center, and patients treated following standard care or through "drugs-in-special-situation" request in Spain prospectively. Patients are not treated to be included in the study, only the follow-up variables are taken into account

Primary outcome measures

Outcome measure	Measure description	Time frame
Points of improvement in the single item dry eye questionnaire (SIDEQ)	The SIDEQ gives a 0–4 score about the presence of dryness, foreign body sensation, burning/stinging, pain, itching, sensitivity to light, and blurred vision, where higher scores indicate higher intensity of the symptom (maximum score: 28). One of the primary outcomes for this work is the improvement in any of the four questionnaires shown here	6 months
Points of improvement about the ocular surface disease index (OSDI)	The OSDI evaluates ocular surface symptoms with 12 questions, and scores >12 indicate abnormal symptomatology, and >32 means severe symptoms (maximum score 100). One of the primary outcomes for this work is the improvement in any of the four questionnaires shown here	6 months
Points of improvement in the national eye institute 25-item visual function questionnaire (NEI-VFQ25)	The visual function-related aspects of the quality of life were evaluated with the NEI-VFQ25, where higher scores on a 0–100 scale indicate better function. One of the primary outcomes for this work is the improvement in any of the four questionnaires shown here	6 months
Points of improvement in the change in dry eye symptoms questionnaire (CDES-Q)	The CDES-Q evaluates ocular surface symptoms with two questions, CDES-Q1 and CDES-Q2. CDES-Q1 evaluates the change in symptoms occurred since the previous examination (same, better, or worse). CDES-Q2 evaluates the magnitude of the change in a scale from 0 to 10, where 0 means "slightly better/worse" and 10 means "extremely better/worse." One of the primary outcomes for this work is the improvement in any of the four questionnaires shown here	6 months
Percentage of improvement in corneal conjunctivalization	Percentage of corneal area with conjunctivalization (superficial opacity and/or superficial neovascularization) will be measured	6 months
Complete absence of persistent epithelial defects	Corneal fluorescein staining has to rule out epithelial breakdown	6 months

Sponsor
Instituto Universitario de Oftalmobiología Aplicada (Institute of Applied Ophthalmobiology)—IOBA

Collaborators
- Citospin
- Red de Terapia Celular
- University of Valladolid

Investigators
No information provided

General Publications
No publications available.

United States/Australia/Brazil/Canada/Egypt

Active, Not Recruiting

Combination Chemotherapy, Autologous Stem Cell Transplant, and/or Radiation Therapy in Treating Young Patients With Extraocular Retinoblastoma

ClinicalTrials.gov ID NCT00554788

Sponsor Children's Oncology Group
Information provided by Children's Oncology Group (Responsible Party)
Last Update Posted 2022-07-12

Brief Summary
This phase III trial is studying the side effects and how well giving combination chemotherapy together with autologous stem cell transplant and/or radiation therapy works in treating young patients with extraocular retinoblastoma. Giving chemotherapy before an autologous stem cell transplant stops the growth of tumor cells by stopping them from dividing or killing them. After treatment, stem cells are collected from the patient's blood and/or bone marrow and stored. More chemotherapy is given to prepare the bone marrow for the stem cell transplant. The stem cells are then returned to the patient to replace the blood-forming cells that were destroyed by the chemotherapy. Radiation therapy uses high-energy X-rays to kill tumor cells. Giving radiation therapy after combination chemotherapy and/or autologous stem cell transplant may kill any remaining tumor cells.

Objectives

1. To estimate the proportion of 3 groups of patients with extraocular retinoblastoma (stage 2 and 3: regional extra-ocular disease; stage 4a: disseminated metastatic disease not involving the central nervous system [CNS]; or stage 4b: patients with CNS disease) who achieve long-term event-free survival after treatment with aggressive multimodality therapy compared to historical controls.
2. To estimate the response rate to the induction phase of the regimen. III. To evaluate the toxicities associated with this regimen.

Outline

Induction chemotherapy: Patients receive vincristine intravenously (IV) on days 0, 7, and 14, cisplatin IV over 6 h on day 0, cyclophosphamide IV over 1 h and etoposide IV over 1 h on days 1 and 2, and filgrastim (G-CSF) subcutaneously (SC) beginning on day 3 and continuing until blood counts recover.

Treatment repeats every 21 days for up to 4 courses in the absence of disease progression or unacceptable toxicity.

After completion of induction chemotherapy, patients with stage 2 or 3 disease who have at least a partial response proceed to radiotherapy. Patients with stage 4a or 4b disease who have at least a partial response proceed to high-dose consolidation chemotherapy and autologous stem cell infusion.

Stem cell harvesting (stage 4a or 4b disease only): Peripheral blood stem cells (preferred) or bone marrow cells are collected after at least 1 course of induction chemotherapy.

High-dose consolidation chemotherapy (stage 4a or 4b disease only): Patients receive carboplatin IV over 4 h on days −8 to −6 and thiotepa IV over 3 h and etoposide IV over 3 h on days −5 to −3.

Autologous stem cell infusion (stage 4a or 4b disease only): Patients undergo autologous stem cell infusion on day 0. Patients then receive filgrastim subcutaneously (SC) beginning on day 1 and continuing until blood counts recover.

Radiotherapy: Patients with stage 2 or 3 disease (orbital and/or regional involvement) undergo radiotherapy to sites that were initially involved beginning within 42 days after the start of course 4 of induction chemotherapy. Patients with stage 4a or 4b disease undergo radiotherapy to sites initially involved based on response beginning approximately 42 days after autologous stem cell infusion. Patients with stage 4a disease who achieve a complete response to induction chemotherapy or with less than 5 mm of residual tumor at the time of planned irradiation, or patients with stage 4b disease who achieve a complete response to induction chemotherapy do not undergo radiotherapy.

After completion of study therapy, patients are followed every 3 months for 1 year and then annually thereafter for 9 years.

Official Title
A Trial of Intensive Multi-Modality Therapy for Extra-Ocular Retinoblastoma

Conditions
Extraocular Retinoblastoma

Study Type
Interventional

Enrollment (Actual)
60

Intervention/Treatment
Procedure: Autologous Bone Marrow Transplantation
Procedure: Autologous Hematopoietic Stem Cell Transplantation
Drug: Carboplatin
Drug: Cisplatin
Drug: Cyclophosphamide
Drug: Etoposide
Biological: Filgrastim
Procedure: In Vitro-Treated Peripheral Blood Stem Cell Transplantation
Radiation: Radiation Therapy
Drug: Thiotepa
Drug: Vincristine Sulfate

Phase
Phase 3

Other Study ID Numbers
ARET0321
NCI-2009-00421 (registry identifier) (REGISTRY: CTRP (Clinical Trial Reporting Program))
COG-ARET0321
ARET0321
CDR0000573987
ARET0321 (other identifier) (OTHER: Childrens Oncology Group)
ARET0321 (other identifier) (OTHER: CTEP)
U10CA180886 (U.S. NIH Grant/Contract)
U10CA098543 (U.S. NIH Grant/Contract)

Study Start (Actual)
2008-02-04

Primary Completion (Actual)
2018-06-30

Study Completion (Estimated)
2024-12-31

United States
Birmingham, Alabama, United States, 35233
Children's Hospital of Alabama
Birmingham, Alabama, United States, 35233
University of Alabama at Birmingham Cancer Center

Arizona Locations
Phoenix, Arizona, United States, 85016
Phoenix Childrens Hospital

Arkansas Locations
Little Rock, Arkansas, United States, 72202-3591
Arkansas Children's Hospital
Little Rock, Arkansas, United States, 72205
University of Arkansas for Medical Sciences

California Locations
Downey, California, United States, 90242
Kaiser Permanente Downey Medical Center
Los Angeles, California, United States, 90027
Children's Hospital Los Angeles
Palo Alto, California, United States, 94304
Lucile Packard Children's Hospital Stanford University
San Francisco, California, United States, 94143
UCSF Medical Center-Parnassus
San Francisco, California, United States, 94158
UCSF Medical Center-Mission Bay

Colorado Locations
Aurora, Colorado, United States, 80045
Children's Hospital Colorado
Denver, Colorado, United States, 80218
Rocky Mountain Hospital for Children-Presbyterian Saint Luke's Medical Center

Connecticut Locations
Hartford, Connecticut, United States, 06106
Connecticut Children's Medical Center

Delaware Locations
Wilmington, Delaware, United States, 19803
Alfred I DuPont Hospital for Children

District of Columbia Locations
Washington, District of Columbia, United States, 20010
Children's National Medical Center

Florida Locations
Jacksonville, Florida, United States, 32207
Nemours Children's Clinic-Jacksonville

Miami, Florida, United States, 33136
University of Miami Miller School of Medicine-Sylvester Cancer Center

Orlando, Florida, United States, 32806
Nemours Children's Clinic—Orlando

Orlando, Florida, United States, 32827
Nemours Children's Hospital

Pensacola, Florida, United States, 32504
Nemours Children's Clinic—Pensacola

Saint Petersburg, Florida, United States, 33701
Johns Hopkins All Children's Hospital

Tampa, Florida, United States, 33607
Saint Joseph's Hospital/Children's Hospital-Tampa

Georgia Locations
Atlanta, Georgia, United States, 30322
Children's Healthcare of Atlanta—Egleston

Illinois Locations
Chicago, Illinois, United States, 60611
Lurie Children's Hospital-Chicago
Chicago, Illinois, United States, 60612
University of Illinois

Indiana Locations
Indianapolis, Indiana, United States, 46202
Riley Hospital for Children

Iowa Locations
Iowa City, Iowa, United States, 52242
University of Iowa/Holden Comprehensive Cancer Center

Kentucky Locations
Lexington, Kentucky, United States, 40536
University of Kentucky/Markey Cancer Center

Maryland Locations
Bethesda, Maryland, United States, 20889-5600
Walter Reed National Military Medical Center

Massachusetts Locations
Boston, Massachusetts, United States, 02114
Massachusetts General Hospital Cancer Center
Boston, Massachusetts, United States, 02215
Dana-Farber Cancer Institute

Mississippi Locations
Jackson, Mississippi, United States, 39216
University of Mississippi Medical Center

Missouri Locations
Kansas City, Missouri, United States, 64108
The Childrens Mercy Hospital
Saint Louis, Missouri, United States, 63110
Washington University School of Medicine

Nevada Locations
Las Vegas, Nevada, United States, 89106
Nevada Cancer Research Foundation CCOP
Las Vegas, Nevada, United States, 89109
Sunrise Hospital and Medical Center
Las Vegas, Nevada, United States, 89135
Alliance for Childhood Diseases/Cure 4 the Kids Foundation
Las Vegas, Nevada, United States, 89144
Summerlin Hospital Medical Center

New Jersey Locations
Morristown, New Jersey, United States, 07960
Morristown Medical Center

New York Locations
New York, New York, United States, 10065
Memorial Sloan Kettering Cancer Center
Valhalla, New York, United States, 10595
New York Medical College

North Carolina Locations
Charlotte, North Carolina, United States, 28203
Carolinas Medical Center/Levine Cancer Institute
Charlotte, North Carolina, United States, 28204
Novant Health Presbyterian Medical Center
Durham, North Carolina, United States, 27710
Duke University Medical Center

Ohio Locations
Akron, Ohio, United States, 44308
Children's Hospital Medical Center of Akron
Cincinnati, Ohio, United States, 45229
Cincinnati Children's Hospital Medical Center
Cleveland, Ohio, United States, 44106
Rainbow Babies and Childrens Hospital
Cleveland, Ohio, United States, 44195
Cleveland Clinic Foundation
Dayton, Ohio, United States, 45404
Dayton Children's Hospital

Oklahoma Locations
Oklahoma City, Oklahoma, United States, 73104
University of Oklahoma Health Sciences Center

Pennsylvania Locations
Hershey, Pennsylvania, United States, 17033
Penn State Children's Hospital
Philadelphia, Pennsylvania, United States, 19104
Children's Hospital of Philadelphia
Philadelphia, Pennsylvania, United States, 19104
Childrens Oncology Group
Pittsburgh, Pennsylvania, United States, 15224
Children's Hospital of Pittsburgh of UPMC

Tennessee Locations
Memphis, Tennessee, United States, 38105
St. Jude Children's Research Hospital

Texas Locations
Dallas, Texas, United States, 75230
Medical City Dallas Hospital
Dallas, Texas, United States, 75390
UT Southwestern/Simmons Cancer Center-Dallas
Fort Worth, Texas, United States, 76104
Cook Children's Medical Center
Houston, Texas, United States, 77030
Baylor College of Medicine/Dan L Duncan Comprehensive Cancer Center
Houston, Texas, United States, 77030
M D Anderson Cancer Center
San Antonio, Texas, United States, 78207
Children's Hospital of San Antonio
San Antonio, Texas, United States, 78229
Methodist Children's Hospital of South Texas
San Antonio, Texas, United States, 78229
University of Texas Health Science Center at San Antonio

Virginia Locations
Richmond, Virginia, United States, 23298
Virginia Commonwealth University/Massey Cancer Center

Wisconsin Locations
Green Bay, Wisconsin, United States, 54301
Saint Vincent Hospital Cancer Center Green Bay
Milwaukee, Wisconsin, United States, 53226
Children's Hospital of Wisconsin

Argentina
Buenos Aires, Argentina, C1245AAL
Hospital de Pediatria Juan P Garrahan

Australia
New South Wales Locations
Westmead, New South Wales, Australia, 2145
The Children's Hospital at Westmead

Western Australia Locations
Perth, Western Australia, Australia, 6008
Princess Margaret Hospital for Children

Brazil
Sao Paulo, Brazil, 04023-062
Instituto De Oncologia Pediatrica

Canada
Nova Scotia Locations
Halifax, Nova Scotia, Canada, B3K 6R8
IWK Health Centre

Quebec Locations
Montreal, Quebec, Canada, H3T 1C5
Centre Hospitalier Universitaire Sainte-Justine

Egypt
Cairo Locations
El Saida Zenab, Cairo, Egypt, 11787
Children's Cancer Hospital

Eligibility Criteria
Ages Eligible for Study
Up to 10 years (child)

Accepts Healthy Volunteer
No

Sexes Eligible for Study
All

Inclusion Criteria
- Patients must have histologic or cytologic verification of extra-ocular retinoblastoma; extra-ocular disease includes orbital disease, optic nerve involvement at the surgical margin, regional nodal disease, and/or overt distant metastatic disease (at sites such as bone, bone marrow, liver, and/or the central nervous system); patients with trilateral retinoblastoma will also be included in this protocol
  - Patients with a CNS lesion consistent with trilateral or stage 4b disease may be enrolled without tissue confirmation if (1) unequivocal leptomeningeal disease is present on brain or spine magnetic resonance imaging (MRI) scan and/or (2) the primary tumor is at least 2 cm in diameter, predominantly solid, and demonstrates enhancement on the post-gadolinium images; however, even in such cases surgery should be given serious consideration

- Patients must have a performance status corresponding to Eastern Cooperative Oncology Group (ECOG) scores of 0, 1, or 2; use Karnofsky for patients >16 years of age and Lansky for patients ≤16 years of age
- No prior chemotherapy or radiotherapy for the extra-ocular retinoblastoma may have been administered prior to entering this study; prior treatment (chemotherapy and/or radiation therapy) for intraocular retinoblastoma is permissible
- Peripheral absolute neutrophil count (ANC) ≥750/uL
  - If the ANC and/or platelet count are not adequate, but due to bone marrow metastatic disease, these criteria will be waived
- Platelet count ≥75,000/uL (transfusion independent)
  - If the ANC and/or platelet count are not adequate, but due to bone marrow metastatic disease, these criteria will be waived
- Creatinine clearance OR radioisotope glomerular filtration rate ≥70 mL/min/1.73 m^2 or a serum creatinine based on age/gender as follows:
  - 0.4 mg/dL (1 month to <6 months of age)
  - 0.5 mg/dL (6 months to <1 year of age)
  - 0.6 mg/dL (1 year to <2 years of age)
  - 0.8 mg/dL (2 years to <6 years of age)
  - 1.0 mg/dL (6 years to <10 years of age)
  - 1.2 mg/dL (10 years to <13 years of age)
  - 1.5 mg/dL (male) or 1.4 mg/dL (female) (13 years to <16 years of age)
  - 1.7 mg/dL (male) or 1.4 mg/dL (female) (≥16 years of age)
- Total bilirubin ≤1.5 times upper limit of normal (ULN)
- Serum glutamic oxaloacetic transaminase (SGOT) (aspartate aminotransferase [AST]) or serum glutamate pyruvate transaminase (SGPT) (alanine aminotransferase [ALT]) <2.5 × upper limit of normal (ULN)
- All patients and/or their parents or legal guardians must sign a written informed consent
- All institutional, Food and Drug Administration (FDA), and National Cancer Institute (NCI) for human studies must be met

Primary purpose: treatment
Allocation: non-randomized
Interventional model: single-group assignment
Masking: none (open label)

Arms and interventions

Participant group/arm	Intervention/treatment
Experimental: Treatment (chemotherapy, radiotherapy, autologous SCI) Induction: Patients receive vincristine IV; cisplatin IV; cyclophosphamide IV; and G-CSF SC beginning on day 3 and continuing until blood counts recover CONSOLIDATION (stage 4a or 4b disease only): Patients receive carboplatin IV; thiotepa IV; and etoposide IV Autologous stem cell infusion (stage 4a or 4b disease only): Patients undergo autologous stem cell infusion on day 0 and receive G-CSF SC beginning on day 1 and continuing until blood counts recover Radiotherapy: Patients with stage 2 or 3 disease (orbital and/or regional involvement) undergo radiotherapy to sites that were initially involved beginning within 42 days after the start of course 4 of induction chemotherapy. Patients with stage 4a or 4b disease undergo radiotherapy to sites initially involved based on response beginning approximately 42 days after autologous stem cell infusion	Procedure: Autologous bone marrow transplantation • Undergo peripheral blood stem cell or bone marrow transplant • Other Names: – ABMT – Autologous bone marrow transplant – Autologous marrow transplantation Procedure: Autologous hematopoietic stem cell transplantation • Undergo peripheral blood stem cell or bone marrow transplant • Other Names: – Autologous hematopoietic cell transplantation – Autologous stem cell transplantation Drug: Carboplatin • Given IV • Other Names: – Blastocarb – Carboplat – Carboplatin Hexal – Carboplatino – Carbosin – Carbosol – Carbotec – CBDCA – Displata – Ercar – JM-8 – Nealorin – Novoplatinum – Paraplatin – Paraplatin AQ – Paraplatine – Platinwas – Ribocarbo Drug: Cisplatin • Given IV • Other Names: – Abiplatin – Blastolem – Briplatin – CDDP – Cis-diammine-dichloroplatinum – Cis-diamminedichloridoplatinum – Cis-diamminedichloro platinum (II) – Cis-diamminedichloroplatinum – Cis-dichloroammine platinum (II) – Cis-platinous diamine dichloride – Cis-platinum – Cis-platinum II – Cis-platinum II diamine dichloride – Cismaplat – Cisplatina – Cisplatinum

	– Cisplatyl – Citoplatino – Citosin – Cysplatyna – DDP – Lederplatin – Metaplatin – Neoplatin – Peyrone's Chloride – Peyrone's Salt – Placis – Plastistil – Platamine – Platiblastin – Platiblastin-S – Platinex – Platinol – Platinol-AQ – Platinol-AQ – Platinol-AQ VHA Plus – Platinoxan – Platinum – Platinum Diamminodichloride – Platiran – Platistin – Platosin Drug: Cyclophosphamide • Given IV • Other Names: – (-)-Cyclophosphamide – 2H-1,3,2-Oxazaphosphorine, 2-[bis (2-chloroethyl)amino]tetrahydro-, 2-oxide, monohydrate – Carloxan – Ciclofosfamida – Ciclofosfamide – Cicloxal – Clafen – Claphene – CP monohydrate – CTX – CYCLO-cell – Cycloblastin – Cycloblastine – Cyclophospham – Cyclophosphamid monohydrate – Cyclophosphamidum – Cyclophosphan – Cyclophosphane – Cyclophosphanum – Cyclostin – Cyclostine – Cytophosphan

	– Cytophosphane – Cytoxan – Fosfaseron – Genoxal – Genuxal – Ledoxina – Mitoxan – Neosar – Revimmune – Syklofosfamid – WR-138719 Drug: Etoposide • Given IV • Other names: – Demethyl epipodophyllotoxin ethylidine glucoside – EPEG – Lastet – Toposar – Vepesid – VP 16-213 – VP-16 – VP-16-213 Biological: Filgrastim • Given SC • Other names: – Filgrastim, license holder unspecified – G-CSF – Neupogen – r-metHuG-CSF – Recombinant Methionyl Human Granulocyte Colony Stimulating Factor – rG-CSF – Tevagrastim Procedure: In Vitro-Treated Peripheral Blood Stem Cell Transplantation • Undergo peripheral blood stem cell or bone marrow transplant • Other Names: – In vitro-treated PBPC transplantation – In vitro-treated peripheral blood progenitor cell transplantation Radiation: Radiation therapy • Undergo radiotherapy • Other names: – Cancer radiotherapy – Irradiate – Irradiated – Irradiation – Radiation – Radiotherapeutics – Radiotherapy – RT – Therapy, radiation Drug: Thiotepa • Given IV • Other names:

	– 1,1′,1″-Phosphinothioylidynetrisaziridine – Girostan – N,N′, N″-Triethylenethiophosphoramide – Oncotiotepa – STEPA – Tepadina – TESPA – Tespamin – Tespamine – Thio-Tepa – Thiofosfamide – Thiofozil – Thiophosphamide – Thiophosphoramide – Thiotef – Tifosyl – Tio TEF – Tio-tef – Triethylene thiophosphoramide – Triethylenethiophosphoramide – Tris(1-aziridinyl)phosphine sulfide – TSPA – WR 45312 Drug: Vincristine sulfate • Given IV • Other names: – Kyocristine – Leurocristine sulfate – Leurocristine, sulfate – Oncovin – Vincasar – Vincosid – Vincrex – Vincristine, sulfate

Primary outcome measures

Outcome measure	Measure description	Time frame
Event-free Survival (EFS)	The probability of surviving patients who did not experience events at 1 year following enrollment. An event is defined as relapse, second malignancy, or death from any cause	At 1 year

Secondary outcome measures

Outcome measure	Measure description	Time frame
Response rate to the induction phase of the regimen	This study used a modified version of the international criteria for neuroblastoma response. The response rate to the induction phase of the regimen and a corresponding 95% confidence interval will be calculated for all strata combined	12 weeks after participant received the first dose

Percentage of participants with adverse events as assessed by the common terminology criteria for adverse events (CTCAE) version 4.0	Grade 3 and higher toxicities will be descriptively summarized	Up to 30 days after completion of study treatment

Sponsor
Children's Oncology Group

Collaborators
- National Cancer Institute (NCI)

Investigators
- Principal Investigator: Ira J Dunkel, Children's Oncology Group

General Publications
No publications available.

Active, Not Recruiting

Phase 1 Study of Intravitreal Autologous CD34+ Stem Cell Therapy for Retinitis Pigmentosa (BMSCRP1)

ClinicalTrials.gov ID NCT04925687

Sponsor University of California, Davis
Information provided by Susanna Park, MD, PhD, University of California, Davis (Responsible Party)
Last Update Posted 2023-04-21

Brief Summary
In this Phase 1, open-labeled prospective study, one eye of each participant with vision loss from retinitis pigmentosa will be administered intravitreal injection of autologous CD34+ stem cells harvested from bone marrow. Each participant will be examined serially for 6 months after study injection to determine safety and feasibility of this intervention.

Official Title
Phase 1 Study of the Feasibility and Safety of Intravitreal Autologous Bone Marrow CD34+ Stem Cell Therapy for Eyes With Vision Loss From Retinitis Pigmentosa

Conditions
Retinitis pigmentosa

Study Type
Interventional

Enrollment (Estimated)
4

Intervention/Treatment
Biological: Intravitreal autologous CD34+ cells

Phase
Phase 1

Other Study ID Numbers
1743714-2

Study Start (Actual)
2021-06-01

Primary Completion (Estimated)
2023-11-30

Study Completion (Estimated)
2023-11-30

United States
Sacramento, California, United States, 95817
University of California Davis

Eligibility Criteria
Ages Eligible for Study
18 years and older (adult, older adult)

Accepts Healthy Volunteer
No

Sexes Eligible for Study
All

Inclusion Criteria
- Age >18 years of age
- Visual Acuity at Presentation of 20/100 to Counting Fingers or constriction of peripheral visual field 360° to <10° by perimetry if BCVA is better than 20/100, duration of vision loss >3 months
- Vision loss due to retinitis pigmentosa
- The study eye has the worse visual acuity or peripheral constriction.
- Able and willing to sign consent
- Able to keep follow-up appointments for at least 6 months

Exclusion Criteria
- Other concurrent optic nerve or retinal disease in study eye affecting vision
- History of active retinal or choroidal neovascularization or macular edema in study eye requiring ongoing intravitreal or periocular therapy in study eye
- Active eye or systemic infection

- Significant media opacity in the study eye precluding view of the fundus for examination, photography, or optical coherence tomography (OCT)
- Other cause for vision loss in the study eye
- History of coagulopathy or other hematologic abnormality
- Use of coumadin within 3 days of enrollment
- Concurrent immunosuppressive therapy
- History of allergy to fluorescein dye

Primary purpose: treatment
Allocation: N/A
Interventional model: single-group assignment
Masking: none (open label)

Arms and interventions

Participant group/arm	Intervention/treatment
Experimental: Intravitreal autologous CD34+ cells Intravitreal injection of autologous CD34+ cells harvested from bone marrow under GMP conditions	Biological: Intravitreal autologous CD34+ cells • Autologous CD34+ cells harvested from bone marrow injected intravitreal

Primary outcome measures

Outcome measure	Measure description	Time frame
Safety of intravitreal injection of autologous CD34+ cells	Number and severity of ocular and systemic adverse events	6 months
Feasibility of intravitreal injection of autologous CD34+ cells	Number of CD34+ cells harvested for intravitreal injection	Baseline

Secondary outcome measures

Outcome measure	Measure description	Time frame
Best-corrected visual acuity	Change in number of letters read in study eye from baseline	6 months
Best-corrected visual acuity	Change in number of letters read in study eye from baseline	1 month
Visual field	Change in visual field area in study eye from baseline	6 months
Visual field	Change in visual field area in study eye from baseline	1 month
Electroretinography	Change in a- and b-wave amplitude in study eye from baseline	6 months
Electroretinography	Change in a- and b-wave amplitude in study eye from baseline	1 month
National eye institute vision questionnaire	Change in vision questionnaire compared to baseline	6 months
National eye institute vision questionnaire	Change in vision questionnaire compared to baseline	1 month
Microperimetry	Change in average threshold from baseline	6 months

Microperimetry	Change percent reduced sensitivity from baseline	6 months
Microperimetry	Change in average threshold from baseline	1 month
Microperimetry	Change in percent reduced sensitivity from baseline	1 month

Sponsor
University of California, Davis

Collaborators
- Cures Within Reach
- Retina Society

Investigators
- Principal Investigator: Susanna Park, MD PhD, University of California, Davis

General Publications
No publications available.

Active, Not Recruiting

Treatment of Central Retinal Vein Occlusion Using Stem Cell Study (TRUST)

ClinicalTrials.gov ID NCT03981549

Sponsor The Emmes Company, LLC
Information provided by The Emmes Company, LLC (Responsible Party)
Last Update Posted 2023-04-19

Brief Summary
This study evaluates whether intravitreal autologous CD34+ stem cell therapy is safe, feasible, and potentially beneficial in eyes with vision loss from central retinal vein occlusion (CRVO). Half of the participants will receive immediate cellular therapy followed by sham therapy 6 months later, while the other half will receive immediate sham therapy followed by cellular therapy 6 months later. Participants will be followed for a total of 1 year.

Detailed Description
The goal of this phase I/II prospective, randomized, sham-controlled, double-masked clinical trial is to determine whether intravitreal autologous CD34+ stem cell therapy is safe, feasible, and potentially beneficial in eyes with vision loss from central retinal vein occlusion (CRVO). Retinal vein occlusion (RVO) is a leading retinal vascular cause of vision loss in the elderly. CD34+ stem cells in human bone marrow are mobilized into the circulation in response to tissue ischemia for tissue revascularization and repair. Since local delivery of CD34+ stem cells benefits

ischemic tissue, intravitreal delivery of CD34+ stem cells may benefit vision and retinal ischemia in eyes with RVO. A pilot clinical trial has shown no major safety or feasibility concerns using intravitreal autologous CD34+ bone marrow stem cells. In this proposed expanded phase I/II study, 20 participants (20 eyes) with persistent vision loss from CRVO will be enrolled and followed for 1 year.

Participants will be randomized 1:1 to immediate cell therapy/deferred sham therapy or immediate sham therapy/deferred cell therapy. At month 6, the cell-treated eye will receive sham treatment and the sham-treated eye will get cell therapy. The cellular therapy involves bone marrow aspiration, isolation of CD34+ cells from the aspirate under good manufacturing practice (GMP) conditions, and intravitreal injection of isolated CD34+ cells. The sham therapy involves a sham bone marrow aspiration with penetration of the skin but no penetration of the bone and a sham intravitreal injection without penetrating the eye. The participant, examining ophthalmologist, visual acuity examiner, photographers and OCT, perimetry, and electroretinography (ERG) technicians will remain masked to study treatment assignment for study duration. A comprehensive eye examination with ETDRS best-corrected visual acuity, optical coherence tomography (OCT) and OCT angiography (OCTA), autofluorescence, fundus photography, fluorescein angiography, microperimetry, and electroretinography will be performed at baseline and serially. A subset of participants with good fixation on microperimetry and clear media on exam and commercial-grade OCTA and who give consent will have ultra-high resolution cellular retinal imaging using research-grade OCT and OCTA and adaptive optics-OCT at baseline. Participants with high-quality images will have repeat imaging at 1 month after stem cell treatment, with at least 2 of the participants randomized to the deferred cellular therapy arm also having imaging 1 month after sham therapy. For all participants in whom at least 1.5 million CD34+ cells are harvested, about 200,000 cells will be set aside for post-release flow cytometry characterization to determine the composition of the CD34+ enriched final product in terms of hematopoietic versus angiogenic stem cells based on cell surface markers (i.e., CD133(+)/CD45(+)/CD34(+) vs CD31(+)/VEGFR-2(+)/CD45(-)/CD34(+)). The long-term objective is to determine whether intravitreal autologous CD34+ cell therapy can minimize, or reverse vision loss associated with retinal ischemia without compromising safety.

Official Title

Phase I/II Randomized, Prospective, Double-masked, Sham-controlled Study of Intravitreal Autologous Bone Marrow CD34+ Stem Cell Therapy for Central Retinal Vein Occlusion

Conditions

Central Retinal Vein Occlusion

Study Type

Interventional

Enrollment (Actual)

16

Intervention/Treatment
Biological: Autologous Bone Marrow CD34+ Stem Cells
Biological: Sham Therapy

Phase
Phase 1 Phase 2

Other Study ID Numbers
1UG1EY026876-01A1 (U.S. NIH Grant/Contract)
1UG1EY028517-01 (U.S. NIH Grant/Contract)

Study Start (Actual)
2019-10-22

Primary Completion (Estimated)
2023-12

Study Completion (Estimated)
2023-12

United States
Sacramento, California, United States, 95817
Department of Ophthalmology & Vision Science, University of California Davis Eye Center

Eligibility Criteria
Ages Eligible for Study
18 years and older (adult, older adult)

Accepts Healthy Volunteer
No

Sexes Eligible for Study
All

Description

Inclusion Criteria for the Study Eye
- Clinical diagnosis of central retinal vein occlusion (CRVO) confirmed by review of medical records and screening assessment.
- Best-corrected visual acuity (BCVA) obtained during the screening period is in the range of 20/40+ to 20/400− (ETDRS letter score in the range of 18–73, inclusive).
- Duration of vision loss from CRVO ≥6 months to ≤42 months.

Exclusion Criteria for the Study Eye
- Previous eye treatment with intravitreal or periocular steroids, laser, or intraocular surgery within 6 months prior to enrollment (i.e., date ICF signed) or treatment expected to be given during the study period.
- For eyes requiring treatment to prevent recurrent macular edema, ongoing intravitreal anti-VEGF treatment is expected to be given at an interval < every 8

weeks during the study period or anti-VEGF therapy was started less than 24 weeks prior to informed consent.
- History of concurrent ocular herpes infection.
- Active non-herpetic eye infection diagnosed within 8 weeks from enrollment (i.e., data informed consent form (ICF) signed).
- Glaucoma requiring treatment with more than 2 medications, laser or intraocular surgery.
- Active uveitis or history of recurrent uveitis or uveitis involving the posterior segment.
- Presence of cataract that is impairing vision.
- Presence of lens or lens implant subluxation.
- History of ocular trauma that is currently impairing vision.
- Any concurrent optic nerve or retinal disease that is visually significant or likely to progress to visual significance during the 1-year study follow-up. The excluded eyes include eyes with AREDS category 2 to 3 age-related macular degeneration (AMD) with foveal involvement of drusen or RPE changes, and any AREDS category 4 AMD eyes. For eyes with ERM, the excluded eyes include eyes with OCT evidence of foveal deformation. For optic nerve disease, eyes with any associated visual field deficit or history of associated CNVM are excluded. For glaucoma eyes, eyes requiring glaucoma laser trabeculectomy or glaucoma surgery to maintain IOP are excluded.
- Active retinal or iris neovascularization.
- Macular edema requiring ongoing therapy or where treatment is expected during the study period, with the exception of anti-VEGF treatment given at an interval of 8 weeks or greater.
- Significant media opacity precluding view of the fundus for examination, photography, or optical coherence tomography (OCT) including cataract and vitreous haze.
- High myopia ($\geq$9 diopters).
- Amblyopia.
- Other cause contributing to vision loss at screening.
- History of any of the following procedures: corneal transplant, glaucoma surgery, or intraocular silicone oil.

Participant-Level Inclusion Criteria
- Age $\geq$18 years.
- Female participants of childbearing potential must not be pregnant or breastfeeding and have a negative urine pregnancy test within 14 days prior to sham injection and/or CD34+ cell injection.
- Females of childbearing potential must have had a hysterectomy, be completely abstinent from intercourse or must agree to practice effective contraception for the duration of the study. Acceptable methods of contraception include hormonal contraception, intrauterine device, barrier methods (diaphragm, condom) with spermicide, or surgical sterilization (tubal ligation).
- Able and willing to sign informed consent.

- Able to keep follow-up appointments for at least 12 months as determined by the investigator.

Participant-Level Exclusion Criteria
- Concurrent treatment with an investigational drug or device.
- Concurrent use of systemic immunosuppressive therapy or history of use within 3 months prior to enrollment (i.e., date ICF signed).
- Concurrent use of anticoagulation therapy except for aspirin without an acceptable safe stopping plan for study treatments.
- Known history of coagulopathy or other hematologic abnormality that may put participant at risk for bleeding or infection or raise concerns about quality or quantity of CD34+ cells isolated.
- History of allergy to fluorescein dye.
- Participant who has had a prior or concomitant malignancy with the exception of the following: 1) adequately treated basal or squamous cell carcinoma of the skin or 2) any other malignancy from which the patient has remained disease free for more than 5 years.
- Current active systemic infection as evidenced by fever greater than 100.4 or any evidence of systemic infection as determined by the study physician.
- Any diagnosis of active infection or vaccination within 8 weeks of study treatment.
- Diabetes mellitus with known systemic complications by self-report or physician-determined by medical history or examination.
- History of prior radiotherapy to head/neck area.
- Poorly controlled hypertension with systolic >180 or diastolic >95.
- Serious medical or psychiatric condition that, in the opinion of the Investigator, would make study participation hazardous to the participant or compromise study findings or would prevent the participant from completing the study.
- Any physical characteristic that precludes ability to perform study diagnostic testing.

Design Details

Primary Purpose: Treatment

Allocation: Randomized

Interventional Model: Parallel Assignment

Masking: Triple (Participant Investigator Outcomes Assessor)

Masking Description: The participant, examining ophthalmologist, visual acuity examiner, photographers and OCT, perimetry, and ERG technicians will remain masked to study treatment assignment for study duration.

Arms and interventions

Participant group/arm	Intervention/treatment
Active comparator: Immediate cellular therapy/deferred sham therapy At baseline: Bone marrow aspiration followed by intravitreal injection of CD34+ cells At 6 months: Sham bone marrow aspiration and sham intravitreal injection	Biological: Autologous bone marrow CD34+ stem cells • Single intravitreal injection of autologous bone marrow CD34+ stem cells. The number of cells to be injected per eye will range from 800,000 to 10 million, depending on the yield of the bone marrow aspiration and the isolation procedure Biological: Sham therapy • Sham bone marrow aspiration procedure that penetrates the skin, but does penetrate the bone followed by sham intravitreal injection without penetration of the eye
Sham comparator: Immediate Sham therapy/deferred cellular therapy At baseline: Sham bone marrow aspiration followed by sham intravitreal injection At 6 months: Bone marrow aspiration followed by intravitreal injection of CD34+ cells	Biological: autologous bone marrow CD34+ stem cells • Single intravitreal injection of autologous bone marrow CD34+ stem cells. The number of cells to be injected per eye will range from 800,000 to 10 million, depending on the yield of the bone marrow aspiration and the isolation procedure Biological: Sham Therapy • Sham bone marrow aspiration procedure that penetrates the skin, but does penetrate the bone followed by sham intravitreal injection without penetration of the eye

Primary outcome measures

Outcome measure	Measure description	Time frame
Incidence and severity of ocular and systemic adverse events	The primary safety outcome will be assessed at month 6. Adverse events (AE) that occur during the first 6 months of the trial will be broken down by whether the AE occurred following the sham or cellular treatment to assess differences in the adverse event experience between the cellular and sham therapies	Months 0–6
Feasibility of the stem cell therapy	Number of CD34+ cells isolated from the bone marrow aspirate and number of cells injected into the eye	Day 0 and Day 182

Secondary outcome measures

Outcome measure	Measure description	Time frame
Change in early treatment diabetic retinopathy study (ETDRS) visual acuity letter score	Mean change from baseline of visual acuity letter score at Month 6. The measure is calculated by subtracting the baseline visual acuity letter score from the month 6 visual acuity letter score. The participant is refracted for best-corrected vision, and then reads single letters from the ETDRS charts using a visual acuity light box at a 4 m distance (or 1 m for participants with sufficiently reduced vision) according to a specific algorithm. A letter score is provided that ranges from 0 (unable to ready any letters) to 100. A visual acuity letter score of 85 corresponds to a visual acuity of 20/20 as a Snellen equivalent	Months 0–6
Change in % reduced sensitivity	Mean change from baseline of % reduced sensitivity at Month 6 as measured by microperimetry	Months 0–6
Change in average threshold	Mean change from baseline of average threshold (dB) at Month 6 as measured by microperimetry	Months 0–6
Change in percent normal amplitude for a-wave under scotopic conditions	Mean change from baseline of percent of normal amplitude for a-wave under scotopic conditions at Month 6 as measured by full-field ERG	Months 0–6
Change in percent normal amplitude for b-wave under scotopic conditions	Mean change from baseline of percent of normal amplitude for b-wave under scotopic conditions at Month 6 as measured by Full-field ERG	Months 0–6
Change in percent normal latency for a-wave under scotopic conditions	Mean change from baseline of percent normal latency for a-wave under scotopic conditions at Month 6 as measured by Full-field ERG	Months 0–6
Change in percent normal latency for b-wave under scotopic conditions	Mean change from baseline of percent normal latency for b-wave under scotopic conditions at Month 6 as measured by Full-field ERG	Months 0–6
Change in percent normal amplitude for a-wave under photopic conditions	Mean change from baseline of percent normal amplitude for a-wave under photopic conditions at Month 6 as measured by full-field ERG	Months 0–6
Change in percent normal amplitude for b-wave under photopic conditions	Mean change from baseline of percent normal amplitude for b-wave under photopic conditions at Month 6 as measured by full-field ERG	Months 0–6
Change in percent normal latency for a-wave under photopic conditions	Mean change from baseline of percent normal latency for a-wave under photopic conditions at Month 6 as measured by full-field ERG	Months 0–6
Change in percent normal latency for b-wave under photopic conditions	Mean change from baseline of percent normal latency for b-wave under photopic conditions at Month 6 as measured by full-field ERG	Months 0–6
Change in percent normal flicker amplitude	Mean change from baseline in percent normal flicker amplitude at Month 6 as measured by full-field ERG	Months 0–6
Change in flicker latency trough	Mean change from baseline of latency trough at Month 6 as measured by full-field ERG	Months 0–6

Change in percent normal amplitude for ERG+OP (oscillatory potential) a-wave under scotopic conditions	Mean change from baseline of percent normal amplitude for ERG+OP a-wave under scotopic conditions at Month 6 as measured by full-field ERG	Months 0–6
Change in percent normal latency of ERG+OP a-wave under scotopic conditions	Mean change from baseline of percent normal latency of ERG+OP a-wave under scotopic conditions at Month 6 as measured by full-field ERG	Months 0-6
Change in percent normal amplitude for ERG+OP b-wave under scotopic conditions	Mean change from baseline of percent normal amplitude for ERG+OP b-wave under scotopic conditions at Month 6 as measured by full-field ERG	Months 0–6
Change in percent normal latency of ERG+OP b-wave under scotopic conditions	Mean change from baseline of percent normal latency of ERG+OP b-wave under scotopic conditions at Month 6 as measured by full-field ERG	Months 0–6
Foveal avascular zone integrity	Number of study eyes in each of the following categories as it relates to the integrity of the foveal avascular zone: intact, questionable, disrupted (<900 microns), disrupted (900–1800 microns), disrupted (>1800 microns), and cannot grade. Measured at Month 6 by fluorescein angiogram	6 months
Change in area of non-perfusion within ETDRS grid	Mean change from baseline of area of non-perfusion within ETDRS grid at Month 6 as measured by fluorescein angiogram	Months 0–6
Change in area of non-perfusion within network grid	Mean change from baseline of area of non-perfusion within Networc grid at Month 6 as measured by fluorescein angiogram	Months 0–6

Sponsor

The Emmes Company, LLC

Collaborators

- National Eye Institute (NEI)
- University of California, Davis

Investigators

- Principal Investigator: Susanna S Park, MD, PhD, University of California, Davis

General Publications

- Park SS, Bauer G, Abedi M, Pontow S, Panorgias A, Jonnal R, Zawadzki RJ, Werner JS, Nolta J. Intravitreal autologous bone marrow CD34+ cell therapy for ischemic and degenerative retinal disorders: preliminary phase 1 clinical trial findings. Invest Ophthalmol Vis Sci. 2014;56(1):81–9. https://doi.org/10.1167/iovs.14-15415.

Active, Not Recruiting

Safety and Tolerability of hRPC in Retinitis Pigmentosa (hRPCRP)

ClinicalTrials.gov ID NCT02464436

Sponsor ReNeuron Limited
Information provided by ReNeuron Limited (Responsible Party)
Last Update Posted 2023-07-06

Brief Summary
hRPC is a cell therapy for retinitis pigmentosa. This is a first-in-human, dose-escalation study in which participants with retinitis pigmentosa will receive a single subretinal injection of hRPC cells in one eye to evaluate safety and tolerability.

Participants will be followed for 2 years to evaluate the safety and tolerability of hRPC Additional testing will seek to establish any preliminary efficacy from hRPC.

Detailed Description
This is a first-in-human open-label phase I/II dose-escalation study in which participants with retinitis pigmentosa will receive a single uniocular subretinal implantation of one of three doses of hRPC.

Treated eyes will be carefully monitored for any ocular or systemic adverse events for 2 years.

Testing will comprise a series of detailed ophthalmic examinations and imaging together with blood testing and systemic evaluations, as necessary.

Ophthalmic testing will also be evaluated for any preliminary efficacy signal.

Official Title
First-in-human Phase I/IIa, Open-Label, Prospective Study of the Safety and Tolerability of Subretinally Transplanted Human Retinal Progenitor Cells (hRPC) in Patients With Retinitis Pigmentosa (RP)

Conditions
Retinitis Pigmentosa

Study Type
Interventional

Enrollment (Actual)
29

Intervention/Treatment
Drug: hRPC

Phase
Phase 1 Phase 2

Other Study ID Numbers
RN03-CP-0001

Study Start
2015-12

Primary Completion (Actual)
2022-06

Study Completion (Estimated)
2023-12

United States
Phoenix, Arizona, United States, 85053
Retinal Research Institute

Massachusetts Locations
Boston, Massachusetts, United States, 02114
Massachusetts Eye and Ear Infirmary

Oregon Locations
Portland, Oregon, United States, 97239
Oregon Health and Science University

No Location

Active, Not Recruiting

Safety and Efficacy Study of OpRegen for Treatment of Advanced Dry-Form Age-Related Macular Degeneration

ClinicalTrials.gov ID NCT02286089

Sponsor Hoffmann-La Roche
Information provided by Hoffmann-La Roche (Responsible Party)
Last Update Posted 2023-05-06

Brief Summary
The main objective of the study is evaluation of the safety and tolerability of OpRegen—Human embryonic stem cell-derived retinal pigment epithelial (RPE) cells. The study will also include initial exploration of the ability of transplanted OpRegen cells to engraft, survive, and moderate disease progression.

Detailed Description
OpRegen® is a cell-based product composed of retinal pigment epithelial (RPE) cells, derived from human embryonic stem cells (hESC) and administered as a cell suspension either in ophthalmic Balanced Salt Solution Plus (BSS Plus) or in

CryoStor® 5 (Thaw-and-Inject, TAI). This is a phase I/IIa, dose escalation, evaluating safety, and tolerability of OpRegen transplantation to patients with progressive dry AMD. The study includes also initial exploration of efficacy.

A total of approximately 24 subjects will be enrolled. The subjects should be 50 years of age and older, with non-neovascular (dry) AMD, who have funduscopic findings of GA in the macula, with absence of additional concomitant ocular disorders. The subjects will be divided into four cohorts, according to their best-corrected visual acuity (BCVA) and administered OpRegen dose.

Official Title
Phase I/IIa Dose Escalation Safety and Efficacy Study of Human Embryonic Stem Cell-Derived Retinal Pigment Epithelium Cells Transplanted Subretinally in Patients With Advanced Dry-Form Age-Related Macular Degeneration (Geographic Atrophy)

Conditions
Age-Related Macular Degeneration

Study Type
Interventional

Enrollment (Actual)
24

Intervention/Treatment
Biological: OpRegen

Phase
Phase 1, Phase 2

Other Study ID Numbers
GR44172

Study Start (Actual)
2015-04-01

Primary Completion (Actual)
2021-12-31

Study Completion (Estimated)
2025-07-01

Eligibility Criteria
Ages Eligible for Study
50 years and older (adult, older adult)

Accepts Healthy Volunteer
No

Sexes Eligible for Study
All

Inclusion Criteria

- Age 50 and older;
- Diagnosis of dry (non-neovascular) age-related macular degeneration in both eyes;
- Funduscopic findings of dry AMD with progressive geographic atrophy in the macula;
- Best-corrected central visual acuity equal or less than 20/200 in cohorts 1–3 and 20/64–20/250 in cohort 4 in the study eye by ETDRS vision testing;
- Vision in the non-operated eye must be better than or equal to that in the operated eye;
- Subjects with sufficiently good health to allow participation in all study-related procedures and complete the study follow-up period (medical records);
- Ability to undergo a vitreoretinal surgical procedure under monitored anesthesia care;
- Blood counts, blood chemistry, coagulation, and urinalysis without abnormal significance;
- Negative for TB (cohort 4), HIV, HBC, and HCV, negative for CMV IgM and EBV IgM;
- No history of malignancy (other than a non-melanoma skin cancer). For cancers in remission for more than 5 years, enrollment is allowed with concurred documented approval of principal investigator and oncologist prior to enrollment;
- Willing to defer all future blood and tissue donation;
- Able to understand and willing to sign informed consent.

Exclusion Criteria

- Evidence of neovascular AMD by history, as well as by clinical exam, fluorescein angiography (FA), or ocular coherence tomography (OCT) at baseline in either eye;
- History or presence of diabetic retinopathy, vascular occlusions, uveitis, Coat's disease, glaucoma, cataract, or media opacity preventing posterior pole visualization or any significant ocular disease other than AMD that has compromised or could compromise vision in the study eye and confound analysis of the primary outcome;
- History of retinal detachment repair in the study eye;
- Axial myopia greater than −6 diopters;
- At least 2 months following cataract removal in the study eye and Yttrium Aluminum Garnet (YAG) laser capsulotomy in the study eye in the past 4 weeks and any other ocular surgery in the study eye in the past 3 months prior to implantation;
- History of cognitive impairments or dementia;
- Contraindication for systemic immunosuppression;
- History of any condition other than AMD associated with choroidal neovascularization in the study eye (e.g., pathologic myopia or presumed ocular histoplasmosis);
- Any type of systemic disease or its treatment, in the opinion of the investigator, including any medical condition (controlled or uncontrolled) that could be

expected to progress, recur, or change to such an extent that it may bias the assessment of the clinical status of the patient to a significant degree or put the patient at special risk;
- Pregnancy or breastfeeding;
- Current participation in another clinical study. Past participation (within 6 months) in any clinical study of a drug administered systemically or to the eye;
- Currently receiving aspirin, aspirin-containing products, and/or any other coagulation-modifying drugs which cannot be discontinued 7 days prior to surgery;
- History of cancer (other than a non-melanoma skin cancer). For cancers cured more than 5 years ago, enrollment is allowed with concurred documented approval of principal investigator and oncologist prior to enrollment.

Primary Purpose: Treatment
Allocation: N/A
Interventional Model: Single-Group Assignment
Masking: None (open label)

Arms and interventions

Participant group/arm	Intervention/treatment
Experimental: OpRegen Up to 12 legally blind subjects with best-corrected visual acuity of 20/200 or less in first three cohorts and 12 subjects with best-corrected visual acuity of 20/64 and 20/250 in fourth cohort	Biological: OpRegen • Targeted dose of 50,000–200,000 cells will be delivered into the subretinal space

Primary outcome measures

Outcome measure	Measure description	Time frame
Percentage of participants with treatment-emergent adverse events	An adverse event is any untoward medical occurrence in a subject administered a pharmaceutical product and which does not necessarily have to have a causal relationship with the treatment. An adverse event can therefore be any unfavorable and unintended sign (including an abnormal laboratory finding, for example), symptom, or disease temporally associated with the use of a pharmaceutical product, whether or not considered related to the pharmaceutical product. Preexisting conditions which worsen during a study are also considered as adverse events. The AE's were graded using National Cancer Institute (NCI) Common Terminology Criteria for Adverse Events (CTCAE) v 3.0	From study start till 12 months following last subject dosed, plus up to 90 days (up to approximately 6.5 years)
Change from baseline in intraocular pressure (IOP)		Baseline, month 12

Secondary outcome measures

Outcome measure	Measure description	Time frame
Change from baseline in geographic atrophy (GA) lesion area	The GA lesion area was based on available fundus autofluorescence (FAF) imaging data by a central reading center	Baseline, month 12
Change from baseline in visual acuity	Change from baseline in visual acuity was measured by retro illuminated ETDRS chart from 4 meters distance. Visual acuity was reported as the number of letters read correctly	Baseline, month 12
Change from baseline in national eye institute visual function questionnaire-25 (NEI VFQ-25) quality of life	The NEI VFQ-25 is a 25-item patient-reported questionnaire. The composite score ranges from 0–100 with the higher score indicating better visual function	Baseline, month 12

Sponsor
Hoffmann-La Roche

Collaborators
No information provided

Investigators
- Study Director: Clinical Trials, Hoffmann-La Roche

General Publications
No publications available.

Results

Description	Change from baseline in visual acuity was measured by retro illuminated ETDRS chart from 4 m distance. Visual acuity was reported as the number of letters read correctly
Time frame	Baseline, month 12
Analysis population description	Efficacy population included all participants who received OpRegen in the study

Arm/group title	Cohort 1	Cohort 2	Cohort 3	Cohort 4
Arm/group description	Participants in cohort 1 with best-corrected visual acuity (BCVA) of 20/200 or less received a single subretinal delivery of the low dose of OpRegen on Day 0	Participants in cohort 2 with BCVA of 20/200 or less received a single subretinal delivery of the high dose of OpRegen on Day 0	Additional participants in cohort 3 with BCVA of 20/200 or less received a single subretinal delivery of the high dose of OpRegen on Day 0	Participants in cohort 4 with BCVA of 20/64 to 20/250 received a single subretinal delivery of the high dose of OpRegen on Day 0

Overall Number of Participants Analyzed	3	3	6	12
Baseline				
Number analyzed	3 participants	3 participants	6 participants	12 participants
Mean (standard deviation)|unit of measure: number of letters[a]	20.0 (18.0)	14.3 (7.2)	29.8 (6.8)	44.8 (7.5)
Change from baseline at month 12				
Number analyzed	3 participants	3 participants	5 participants	12 participants
[a]	15.7 (34.2)	3.0 (8.7)	−1.4 (17.8)	7.6 (8.7)

[a] Mean (standard deviation)|unit of measure: number of letters

Change from Baseline in National Eye Institute Visual Function Questionnaire-25 (NEI VFQ-25) Quality of Life
Type: secondary|time frame: baseline, month 12

Adverse Events

Time Frame
From study start till 12 months following last subject dosed, plus up to 90 days (up to approximately 6.5 years)

Adverse Event Reporting Description
Safety Population included all participants who received OpRegen in the study.

Arm/group title	Cohort 1	Cohort 2	Cohort 3	Cohort 4
Arm/group description	Participants in cohort 1 with best-corrected visual acuity (BCVA) of 20/200 or less received a single subretinal delivery of the low dose of OpRegen on Day 0	Participants in cohort 2 with BCVA of 20/200 or less received a single subretinal delivery of the high dose of OpRegen on Day 0	Additional participants in cohort 3 with BCVA of 20/200 or less received a single subretinal delivery of the high dose of OpRegen on Day 0	Participants in cohort 4 with BCVA of 20/64 to 20/250 received a single subretinal delivery of the high dose of OpRegen on Day 0

Anterior Segment

United States

Active, Not Recruiting

Limbal Stem Cell Deficiency (LSCD) Treatment With Cultivated Stem Cell (CALEC) Graft (CALEC)

ClinicalTrials.gov ID NCT02592330

Sponsor Massachusetts Eye and Ear Infirmary
Information provided by Ula Jurkunas, Massachusetts Eye and Ear Infirmary (Responsible Party)
Last Update Posted 2023-01-26

Brief Summary
The main aim of the study is to determine the safety and feasibility of a cultivated autologous limbal epithelial cell (CALEC) transplantation in the treatment of limbal stem cell deficiency.

Detailed Description
This is an open-label, single-center study to assess safety, feasibility, and efficacy of Cultivated Autologous Limbal Epithelial Cell (CALEC) grafts in 17 patients with unilateral limbal stem cell deficiency (LSCD).

Participants will have a corneal biopsy in their non-diseased eye, which will provide cells for the creation of the CALEC graft. The CALEC will be made at the Good Manufacturing Practice (GMP) Laboratory, Dana Farber Cancer Institute and transported to Mass. Eye and Ear Infirmary for application to the participant's diseased eye during their standard corneal reconstruction procedure. Subjects will be monitored up to month 18 post-transplant to assess for any delayed adverse events of the product (CALEC) or procedure as well as assessment of the durability of the transplant.

Official Title
Safety and Feasibility of Cultivated Autologous Limbal Epithelial Cell Transplantation in the Treatment of Limbal Stem Cell Deficiency (LSCD)

Conditions
Limbal Stem Cell Deficiency

Study Type
Interventional

Enrollment (Estimated)
17

Intervention/Treatment
Procedure: Biopsy to collect limbal epithelial stem cells that will be cultivated into a graft
Biological: Cultivation of Limbal epithelial cells into a graft
Procedure: CALEC Transplant

Phase
Phase 1, Phase 2

Other Study ID Numbers
14-124H

Study Start (Actual)
2016-08-01

Primary Completion (Estimated)
2023-04-01

Study Completion (Estimated)
2023-09-30

United States
Boston, Massachusetts, United States, 02114
Massachusetts Eye and Ear Infirmary

Eligibility Criteria
Ages Eligible for Study
18–89 years (adult, older adult)

Accepts Healthy Volunteer
No

Sexes Eligible for Study
All

Inclusion Criteria
- Male or female participants age 18 to <90 years old at time of enrollment
- Ability of a subject or guardian/legal representative to provide written informed consent and to comply with study assessments for the full duration of the study
- Patients with unilateral limbal stem cell deficiency (LSCD) as determined by conjunctivalization of the cornea defined by fibrovascular pannus more than 2 mm from the limbus for greater than or equal to 6 clock hours
- Additional optional criteria:
  - Lack of limbal palisades of Vogt for greater than or equal to 9 clock hours
  - Goblet cell presence as defined by impression cytologic criteria

Exclusion Criteria
- Corneal or ocular surface infection within 30 days prior to study entry or CALEC transplantation
- Ocular surface malignancy

- Uncontrolled diabetes with most recent HgA1c greater than 8.5%
- Renal failure with eGFR below 60 mL/min per 1.73 m^2
- Aspartate aminotransferase and alanine aminotransferase levels greater than 3 times institutional upper limit of normal
- Total bilirubin greater than 2 times institutional upper limit of normal (except patients with known Gilbert's syndrome)
- Platelet levels less than 100,000 or greater than 450,000/μl
- Hemoglobin levels of less than 11.0 g/dL in men or less than 10.0 g/dL in women
- Prothrombin time greater than 16 s and activated partial thromboplastin time greater than 35 s in patients not taking warfarin and an international normalized ratio greater than 3 in patients taking warfarin
- Inability to tolerate monitored anesthesia
- HIV infection or AIDS
- Active Hepatitis B or C
- Pregnancy (positive test) or lactation
- Participation in another simultaneous medical investigation or trial
- Severe cicatricial eye disease
- Severe dry eye disease as determined by Schirmer's test less than 1 mm in at least one eye.
- Any medical, psychiatric, debilitating disease/disorder or social condition that in the judgment of the investigator would interfere with or serve as a contraindication to adherence to the study protocol or ability to give informed consent.
- Signs of current infection, including fever and current treatment with antibiotics.
- History of allo-limbal transplantation
- Presence of allergy to the CALEC graft or any of the chemical components within its formulation.

Exclusion Based on Donor Eye

- Conjunctivalization of the cornea defined by fibrovascular pannus more than 2 mm from the limbus for greater than or equal to 3 clock hours
- Lack of limbal palisades of Vogt for greater than or equal to 3 clock hours
- History of allo-limbal transplantation

Primary Purpose: Treatment
Allocation: N/A
Interventional Model: Single-Group Assignment
Interventional Model Description: All subjects will receive the study intervention, a stem cell graft cultivated from their own cells.
Masking: None (open label)

Arms and interventions

Participant group/arm	Intervention/treatment
Experimental: Cultivated autologous limbal epithelial cell (CALEC) graft Participants will have a corneal biopsy in their non-diseased eye, which will provide cells for the creation of the CALEC graft. The CALEC will be made at the Good Manufacturing Practice (GMP) Laboratory, Dana Farber Cancer Institute and transported to Mass. Eye and Ear Infirmary for application to the participant's diseased eye during their standard corneal reconstruction procedure	Procedure: Biopsy to collect limbal epithelial stem cells that will be cultivated into a graft • Cultivated autologous limbal epithelial cell (CALEC) therapy utilizes a bio-engineered composite of ex vivo expanded autologous corneal epithelial cells and an FDA-approved amniotic membrane (AmnioGraft®, Bio-Tissue, Inc.) to reconstruct the ocular surface. A small biopsy (2–3 mm^2) from the patient's contralateral eye serves as a source epithelial (stem) cells that are expanded on the amniotic membrane in culture and the resulting product is surgically transplanted onto the cornea after excision of the fibrovascular pannus • Other Names: – Cultivated Autologous Limbal Epithelial Cell (CALEC) Biological: Cultivation of Limbal epithelial cells into a graft • A graft is manufactured for transplant Procedure: CALEC Transplant • Limbal epithelial cells are obtained from the healthy fellow eye and cultivated in a lab for later transplantation into the diseased eye • Other Names: – Conjunctival limbal autograft

Primary outcome measures

Outcome measure	Measure description	Time frame
Safety—The occurrence of ocular infection	Measurement of the incidence of ocular infection	2 years
Safety—The occurrence of corneal perforation	Measurement of the incidence of corneal perforation.	2 years
Safety—The occurrence of graft detachment ≥50%	Measurement of the incidence of graft detachment ≥50%	2 years
Safety—The occurrence adverse events and their relationship to the study intervention	Measurement of the number and severity of adverse events	2 years
Feasibility—obtaining cell growth and maintaining cell viability	Measurement of number of cells in graft at intervals to establish feasibility of manufacturing CALEC for corneal transplantation	2 years
Feasibility—avoiding culture contamination	Measurement of the incidence of culture contamination	2 years

Secondary outcome measures

Outcome measure	Measure description	Time frame
Efficacy—improvement in corneal surface Integrity	Measurement of decreases in frank epithelial defect surface area (success is improvement from baseline to post-op of >75%), and decreases in surface staining (NEI grading scale). Success is improvement from baseline to post-op of >50%	2 years
Efficacy—decrease in neovascularization	Measurement of decrease in neovascular area. Success is decrease in neovascularization from baseline to post-op of >25%	2 years
Efficacy—decrease in subject symptoms	Measurement of change from baseline to post-op in Ocular Surface Disease Index (OSDI) and Symptom Assessment in Dry Eye (SANDE) scores	2 years

Sponsor
Massachusetts Eye and Ear Infirmary

Collaborators
- Dana-Farber Cancer Institute

Investigators
- Principal Investigator: Ula Jurkunas, MD, Massachusetts Eye and Ear Infirmary

General Publications
No publications available.

Active, Not Recruiting

United States/Germany

Allogeneic ABCB5-Positive Limbal Stem Cells for Treatment of LSCD

ClinicalTrials.gov ID NCT03549299

Sponsor RHEACELL GmbH & Co. KG
Information provided by RHEACELL GmbH & Co. KG (Responsible Party)
Last Update Posted 2022-10-07

Brief Summary
The aim of this clinical trial is to investigate the efficacy (by monitoring neovascularization and epithelial defects) of up to four doses of the investigational medicinal product (IMP) LSC2 topically administered on the target eye of patients with LSCD. Further, safety of the IMP during and after application will be investigated (by monitoring adverse events [AEs]).

Detailed Description

This is an interventional, open-label, phase I/IIa clinical trial to investigate the efficacy and safety of up to four doses of the IMP topically administered on the target eye of patients with LSCD. Patients will be treated in up to four ascending dose groups.

The allogeneic investigational product LSC2 contains ABCB5-positive limbal stem cells (from corneal rims of cadaveric donors, expanded ex vivo, isolated, and stored in a donor cell bank).

The IMP will be applied on the target eye. Prior to application, the conjunctival pannus will be removed under general or local anesthesia.

Patients will be followed up for efficacy for 1 year. Efficacy of the IMP will be monitored by assessing neovascularization and epithelial defects.

To assess long-term safety of LSC2 one follow-up visit at Month 24 post-IMP application is included.

Official Title

An Interventional, Open-label, Multicenter Phase I/IIa Clinical Trial to Investigate the Safety and Efficacy of Ascending Doses of Allogeneic ABCB5-positive Limbal Stem Cells (LSC2) for the Treatment of Limbal Stem Cell Deficiency (LSCD)

Conditions

Limbal Stem Cell Deficiency

Study Type

Interventional

Enrollment (Estimated)

16

Intervention/Treatment

Biological: LSC2

Phase

Phase 1, Phase 2

Other Study ID Numbers

LSC2-II-01

Study Start (Actual)

2019-07-04

Primary Completion (Estimated)

2023-09

Study Completion (Estimated)

2023-09

United States

Boston, Massachusetts, United States, 02114

Massachusetts Eye and Ear Infirmary

Germany
Heidelberg, Germany, 69120
Universitäts-Klinikum Heidelberg, Kopfklinik
Jena, Germany, 07747
Universitäts-Klinikum Jena, Augenklinik
Köln, Germany, 50937
Universitäts-Klinikum Köln, Augenklinik
Mainz, Germany, 55131
Universitätsmedizin der Johannes Gutenberg-Universität Mainz, Augenklinik und Poliklinik

Eligibility Criteria
Ages Eligible for Study
18–85 years (adult, older adult)

Accepts Healthy Volunteer
No

Sexes Eligible for Study
All

Inclusion Criteria
- Male or female patients aged 18–85 years
- Patients with secondary bilateral or unilateral LSCD (injury that caused LSCD at least 6 months prior to inclusion)
- Neovascularization: Vessel penetration of at least 2 quadrants, with central cornea involved
- Patients understand the nature of the procedure and are providing written informed consent prior to any clinical trial procedure
- Women of childbearing potential must have a negative blood pregnancy test at Visit 1
- Women of childbearing potential must be willing to use highly effective contraceptive methods during the course of the clinical trial

Exclusion Criteria
- Compromised eyelid mobility and/or symblepharon; patient can be re-screened after appropriate treatment
- Presence of eyelid malposition; patient can be re-screened after appropriate treatment
- Active local ocular or systemic infection and/or inflammation. Patient can be re-screened after infection and/or inflammation is resolved
- Tumor diseases or history of tumor disease
- Active ocular neoplastic disease (exclusion will be based on investigator's assessment)
- Corneal erosion or ulcer is bigger than 4 mm^2; corresponding to less than 95% of continuous corneal epithelium. Patient can be re-screened after erosion or ulcer is resolved ($\leq$4 mm^2)

- Positive for human immunodeficiency virus (HIV) 1 and/or 2 (diagnosed by serologic testing)
- Clinically significant or unstable concurrent disease or other clinical contraindications to stem cell transplantation
- History of glaucoma
- Contraindications to trial-related procedures/substances including
  - The surgical procedure (e.g., removing of the conjunctival pannus)
  - Contact lens complications due to contact lens use in the proposed trial (based on the Efron Grading scale for standard clinical reference for contact lens complications)
  - Tear secretion deficiency determined by Schirmer's test
  - Allergy, sensitivity, or intolerance to components/excipients of the IMP/ per protocol pre-planned concomitant medications
  - Conjunctival scarring with fornix shortening
  - General anesthesia (in case general anesthesia is required) or local anesthesia
  - Immunosuppression (being mandatory concomitant therapy)
- Intraocular pressure (IOP) of ≥30 mmHg
- History or clinical signs of stroke or transient ischemic attacks
- Active or suspected ocular or periocular infections
- Active or suspected intraocular inflammation
- Further clinical contraindications to IMP application (exclusion will be based upon investigator's judgment)
- Current or previous (within 30 days of enrollment) treatment with another IMP, or participation and/or under follow-up in another clinical trial
- Previous participation in this clinical trial (except screening failure due to inclusion criterion 2 and/or exclusion criterion 1 and/or 2 and/or 3 and/or 6)
- Known abuse of alcohol, drugs, or medicinal products
- Patients unwilling or unable to comply with the requirements of the protocol
- Lactating women
- Evidence of any other medical conditions (such as psychiatric illness, physical examination, or laboratory findings) that may interfere with the planned treatment, affect the patient's compliance, or place the patient at high risk of complications related to the treatment
- Employees of the sponsor, or employees or relatives of the investigator

Primary purpose: treatment
Allocation: non-randomized
Interventional model: parallel assignment

Interventional Model Description
Patients will be treated in up to four ascending dose groups. After the first treated patient of the first-dose group (Group A) was followed for 2 weeks, a safety and tolerability assessment of the applied IMP dose will be conducted by an internal committee. The same approach applies to the second subject.

Dose-escalation and recruitment of additional patients into the second-dose group (Group B) will depend on the safety and tolerability of the IMP in Group A. A safety and tolerability assessment will be conducted after the last treated patient of the first-dose group (Group A) was followed for 12 weeks.

The same approach applies to Group C and D.

Masking: none (open label)

Arms and interventions

Participant group/arm	Intervention/treatment
Experimental: LSC2; 7.5×10^4 cells Single dose of LSC2, 7.5×10^4 cells per patient	Biological: LSC2 • Topical application of IMP on target eye • Other Names: – Suspension of ABCB5-positive limbal stem cells in pre-filled syringe
Experimental: LSC2; 3.0×10^5 cells Single dose of LSC2, 3.0×10^5 cells per patient	Biological: LSC2 • Topical application of IMP on target eye • Other Names: – Suspension of ABCB5-positive limbal stem cells in pre-filled syringe
Experimental: LSC2; 8.0×10^5 cells Single dose of LSC2, 8.0×10^5 cells per patient	Biological: LSC2 • Topical application of IMP on target eye • Other Names: – Suspension of ABCB5-positive limbal stem cells in pre-filled syringe
Experimental: LSC2; 1.2×10^6 cells Single dose of LSC2, 1.2×10^6 cells per patient	Biological: LSC2 • Topical application of IMP on target eye • Other Names: – Suspension of ABCB5-positive limbal stem cells in pre-filled syringe

Primary outcome measures

Outcome measure	Measure description	Time frame
Response rate at 12 months after IMP application	Response rate at 12 months after IMP application, where response is defined as: • No or mild corneal neovascularization (no vessel penetration or vessel penetration up to 1 quadrant, without central cornea) and • No or mild epithelial defects (no corneal erosion or ulcer are present (corneal wounds are closed) or minimal superficial staining)	Month 12
Assessment of adverse event (AE) occurrence	All AEs occurring during the clinical trial will be registered, documented, and evaluated	Up to 24 months

Secondary outcome measures

Outcome measure	Measure description	Time frame
Response rate at 3 months after IMP application	Response rate at 3 months after IMP application, where response is defined as: • No or mild corneal neovascularization (no vessel penetration or vessel penetration up to 1 quadrant, without central cornea) and • No or mild epithelial defects (no corneal erosion or ulcer are present (corneal wounds are closed) or minimal superficial staining)	Month 3
Neovascularization	Corneal neovascularization will be assessed as "none" (no blood vessel penetration), "mild" (blood vessel penetration of 1 quadrant, without central cornea involvement), "moderate" (blood vessel penetration of 1 quadrant, with central cornea involvement or blood vessel penetration of 2 or 3 quadrants, with or without central cornea involvement), or "strong" (blood vessel penetration of all quadrants, with or without central cornea involvement)	Baseline, week 1, 2, 3, 4, 5, 6, 7, 12, month 6 and 12
Epithelial defects	Epithelial defects will be assessed as "none" (no corneal erosion or ulcer are present (corneal wounds are closed)) or "mild" (minimal superficial staining) or "moderate" (dense coalescent staining up to 2 mm in diameter) or "strong" (dense coalescent staining greater than 2 mm in diameter) by means of fluorescein staining	Baseline, week 2, 4, 6, month 3, 6 and 12
Ocular symptoms of pain, photophobia, burning	Photophobia and burning will be assessed as "no complaint" (grade 0), "mild" (grade 1), "moderate" (grade 2), or "severe" (grade 3) Pain assessment will be done by the patient using a 11-point numerical rating scale ranging between no pain (zero) and worst imaginable pain (ten)	Baseline, week 2, 4, 6, 12, month 6 and 12
Ocular inflammation	The assessment of inflammation will be categorized as present (yes) or nonexistent (no)	Baseline, week 2, 4, 6, 12, month 6 and 12

Outcome measure	Measure description	Time frame
Corneal opacity	Corneal opacification will be assessed as "none" (no corneal opacification), "mild" (corneal opacification in up to 1 quadrant, without central cornea involvement), "moderate" (corneal opacification in 1 quadrant, with central cornea involvement or corneal opacification in 2 or 3 quadrants, with or without central cornea involvement), or "strong" (corneal opacification in all quadrants, with or without central cornea involvement) Additional densitometric Pentacam® measurements are optional. Opacity values of the following ring-shaped corneal segments will be record: • 0–2 mm, 2–6 mm, 6–10 mm, 10–12 mm (anterior, central and posterior corneal layer for each segment) • 0–2 mm, 2–6 mm, 6–10 mm, 10–12 mm (full thickness average for each segment) • Anterior layer of the complete cornea (0–12 mm) • Central layer of the complete cornea (0–12 mm) • Posterior of the complete cornea (0–12 mm) • Overall cornea (0–12 mm, full thickness)	Baseline, week 2, 4, 6, 12, month 6 and 12
Visual acuity	Clarity of vision	Baseline, week 2, 4, 6, 12, month 6 and 12
Quality-of-life assessment	Quality-of-life (visual function questionnaire-25 [VFQ-25]) questionnaire to be answered	Baseline, week 12, month 12
Physical examination	A full physical examination will be performed at week 12 and abnormal physical examination results will be evaluated and reported as AEs	Baseline, week 12, month 12
Tonometry		Baseline,week 1, 3, 7 and 12
Vital signs: Body temperature at Baseline, week 12, month 12	Body temperature at Baseline, week 12 and month 12 will be evaluated	Baseline, week 12, month 12
Vital signs: Blood pressure at Baseline, week 12, month 12	Blood pressure at Baseline, week 12 and month 12 will be evaluated	Baseline, week 12, month 12
Vital signs: Heart rate at Baseline, week 12, month 12	Heart rate at Baseline, week 12 and month 12 will be evaluated	Baseline, week 12, month 12

Sponsor

RHEACELL GmbH & Co. KG

Collaborators

- FGK Clinical Research GmbH
- Ticeba GmbH
- Granzer Regulatory Consulting & Services

Investigators

- Principal Investigator: Gerd U. Auffarth, Prof.Dr.med., Universitäts-Klinikum Heidelberg, Kopfklinik, Heidelberg, Germany

General Publications

No publications available.

Denmark

Active, Not Recruiting

Treatment with Allogeneic Adipose-Derived Mesenchymal Stem Cells in Patients with Aqueous Deficient Dry Eye Disease (MESADDE)

ClinicalTrials.gov ID NCT03878628

Sponsor Rigshospitalet, Denmark
Information provided by Michael Møller-Hansen, Rigshospitalet, Denmark (Responsible Party)
Last Update Posted 2022-10-27

Brief Summary

The objective of this study is to assess the safety and feasibility of allogeneic adipose tissue-derived mesenchymal stem cells (ASCs) injected into the lacrimal gland in a small group of 7 patients with Aqueous Deficient Dry Eye Disease (ADDE)

Detailed Description

Dry eye disease (DED) is a very common problem seen in patients all over the world. According to an older study the prevalence of DED in a Danish population 30–60 years of age was 11%. Aqueous tear deficient dry eye (ADDE) is a subtype of DED in which the tear production in the lacrimal gland (LG) is impaired. Current treatment of ADDE is only to relieve symptoms as a curative treatment of ADDE does not exist.

Mesenchymal stem cells (MSCs) reside in almost all connective tissues and are multipotent stem cells with the capacity to differentiate into several kinds of tissue. Several studies have shown that MSCs reduce inflammation in various diseases. Adipose tissue-derived MSCs (ASCs) have gained considerable attention, since they are readily available from the abdominal fat where it is most easily collected and expanded. In resting MSCs, MHC class II is not expressed on the surface, which reduces the inherent immunogenicity of the cells. This supreme attribute allows allogeneic MSC transplantation. Treatment with allogeneic MSCs has been investigated in an extensive number of human subjects for various conditions in clinical trials and no documented adverse events related to an anti-donor immune response exist. One potential advantage of treatment with allogeneic cells is the possibility of their use as an "off-the-shelf" therapeutic agent, avoiding the need for tissue collection and culture to delay and increase the cost of treatment. It has also been suggested that the function of autologous MSCs could be impaired in patients with comorbidities or advanced age.

In canines as in humans, the most common cause of ADDE is an immune-mediated inflammatory response targeting the LG. Two studies with injection of allogeneic ASCs from healthy donors in a total of 48 eyes in 27 canines with ADDE have been performed with a significant increase in tear production and no observed adverse events to the treatment.

Studies with injection of ASCs into the human LG have never been conducted. This present study will test the hypothesis that injection of allogeneic ASCs into the LG in patients suffering from ADDE is safe and increases tear production and reduces inflammation resulting in increased ocular comfort.

7 patients with severe ADDE from Dept. of Ophthalmology, Rigshospitalet-Glostrup, will be recruited if they are 1: eligible for the study and 2: sign the informed consent form.

At inclusion, the participants will fill out the Ocular Surface Disease Index (OSDI) questionnaire and undergo an eye examination in the following order:

measurement of tear osmolarity (TearLab™), tear break-up time (TBUT), ocular surface staining according to the Ocular SICCA Grading Score, and Schirmer's I test.

After a maximum of 14 days from screening all participants will receive an injection with ASCs into the lacrimal gland on one eye. If both eyes fulfill the eligibility criteria the most affected eye with the lowest tear production assessed with the Schirmer's I test will be the study eye; the contralateral eye will not be treated but examined according to the same protocol as the study eye at each follow-up.

The product used is CSCC_ASC (22) and the dose injected contains approximately 11 million ASCs per LG in a suspension with a total volume of 0.5 mL.

1 week (±2 days), 4 weeks (±4 days), and 4 months (±7 days) after intervention the participants are followed up with eye examination as described above, OSDI questionnaire, and blood test. At 4 months, the primary outcomes of safety will be evaluated.

End of trial is defined as last participant's last visit (LPLV) at 3 years late follow-up.

Official Title

Treatment With Allogeneic Adipose-derived Mesenchymal Stem Cells in Patients With Aqueous Deficient Dry Eye Disease

Conditions

Dry Eye
Kerato-Conjunctivitis Sicca
Aqueous tear deficiency

Study Type

Interventional

Enrollment (Actual)
7

Intervention/Treatment
Drug: Adipose tissue-derived mesenchymal stem cells

Phase
Early Phase 1

Other Study ID Numbers
001-2018

Study Start (Actual)
2019-10-16

Primary Completion (Actual)
2020-06-30

Study Completion (Estimated)
2023-02-20

Denmark
Copenhagen, DK, Denmark, 2200
Rigshospitalet

Eligibility Criteria
Ages Eligible for Study
18 years and older (adult, older adult)

Accepts Healthy Volunteer
No

Sexes Eligible for Study
All

Description

Inclusion Criteria
- OSDI-score >30
- Schirmer's test 2–5 mm in 5 min
- TBUT <10 s

Exclusion Criteria
- Previously established allergies to Oxybuprocaine or DMSO (rare)
- Reduced immune response (e.g., HIV positive)
- Pregnancy or planned pregnancy within the next 2 years
- Breastfeeding
- Treatment with an anticoagulant that cannot be stopped during the intervention period
- Treatment with systemic medication known to reduce tear production (with an odds ratio >2.0 (3)): anxiolytics, antipsychotics, and inhaled steroids

- Topical treatment with eye drops other than lubricants
- Any other disease/condition judged by the investigator to be grounds for exclusion, such as infection in or around the eye

Design Details

Primary purpose: treatment
Allocation: N/A
Interventional model: single-group assignment
Masking: none (open label)

Arms and interventions

Participant group/arm	Intervention/treatment
Experimental: Adipose tissue-derived mesenchymal stem cells Approximately 11 million ASCs in a 0.5 mL suspension	Drug: Adipose tissue-derived mesenchymal stem cells • CSCC_ASC (22), a ready-to-use suspension containing 22 million adipose tissue-derived mesenchymal stem cells per milliliter

Primary outcome measures

Outcome measure	Measure description	Time frame
Pain at injection site: grade	Grade 1: mild pain, grade 2: moderate pain, grade 3: severe pain	4 months after treatment
Infection at injection site	Grade 1: localized; local intervention indicated, grade 2: oral intervention indicated (antibiotic, antifungal, antiviral), grade 3: intravenously administered (IV) antibiotic, antifungal, or antiviral agent indicated; or operative intervention indicated, grade 4: life-threatening consequences; urgent intervention needed	4 months after treatment
Bleeding at injection site	Grade 1: Mild bleeding; intervention not indicated; Grade 2: Moderate; minimal, local or non-invasive intervention indicated; Grade 3: Severe or medically significant but not immediately life-threatening; hospitalization indicated	4 months after treatment
Eyelid function disorder	Grade 1: Asymptomatic; clinical or diagnostic observations only; intervention not indicated Grade 2: Symptomatic; nonoperative intervention indicated; limiting instrumental ADL Grade 3: Limiting selfcare ADL; operative intervention indicated	4 months after treatment
Periorbital edema	Grade 1: Soft or non-pitting Grade 2: Indurated or pitting edema; topical intervention indicated Grade 3: Edema associated with visual disturbance; increased intraocular pressure, glaucoma or retinal hemorrhage; optic neuritis; diuretics indicated; operative intervention indicated	4 months after treatment

Ocular discomfort	Grade 1: mild discomfort Grade 2: moderate pain Grade 3: disabling pain	4 months after treatment
Flu-like symptoms	Grade 1: Mild flu-like symptoms present Grade 2: Moderate flu-like symptoms, limiting self-care ADL Grade 3: Severe flu-like symptoms, limiting selfcare ADL)	4 months after treatment
Fever	Grade 1: 38.0–39.0 °C; grade 2: >39.0–40.0 °C; grade 3: >40.0 °C for ≤24 h; grade 4: >40.0 °C for >24 h	4 months after treatment

Secondary outcome measures

Outcome measure	Measure description	Time frame
OSDI questionnaire	Change in severity of dry eye symptoms as assessed with the OSDI questionnaire with a score of 0–100, where 0 is "no symptoms" and 100 is "severe symptoms"	4 months after treatment
Schirmer's I test	Change in tear production as evaluated with the Schirmer's I test	4 months after treatment
Tear osmolarity	Change in tear osmolarity measured with TearLab™	4 months after treatment
Ocular SICCA grading score	Change in objective signs of DED as evaluated with the Ocular SICCA Grading Score from 0–12 where 0 is "no staining of the ocular surface" and 12 is "severe staining of the ocular surface"	4 months after treatment
HLA antibodies	Development of anti-HLA antibodies evaluated with Luminex HLA antibody screening	4 months after treatment

Sponsor
Rigshospitalet, Denmark

Collaborators
No information provided

Investigators
- Study Chair: Steffen Heegaard, MD, DMSc, Department of Ophthalmology, Rigshospitalet-Glostrup, University of Copenhagen

General Publications
No publications available.

Results Overview
No Study Results Posted on ClinicalTrials.gov for this Study

France

Active, Not Recruiting

Interventional Study of Implantation of hESC-Derived RPE in Patients with RP Due to Monogenic Mutation

ClinicalTrials.gov ID NCT03963154

Sponsor Centre d'Etude des Cellules Souches
Information provided by Centre d'Etude des Cellules Souches (Responsible Party)
Last Update Posted 2023-01-12

Brief Summary
Phase I/II open-label, safety, tolerability, and preliminary efficacy study of implantation into one eye of hESC-derived RPE (Human Embryonic Stem Cell-Derived Retinal Pigment Epithelium (RPE)) in patients with retinitis pigmentosa due to monogenic mutation.

Study non randomized single-group assignment consisting in 2 sequential cohorts of patients:

- First cohort of 2 patients with very advanced loss of visual acuity (legally blind)
- Second cohort of 10 patients with less advanced loss of visual acuity:

Detailed Description
Monocentric study (Hospital of 15–20 at Paris) of duration of 106 weeks.

At total, 12 evaluable patients will be enrolled and assigned in 2 cohorts as described above in brief summary.

Expected follow-up for one patient is about 64 weeks including 8 weeks of screening and baseline period and 56 weeks of follow-up after implantation of hESC-derived RPE.

After 56 weeks of follow-up, patients will be enrolled in a long-term follow-up study during 4 additional years.

The primary objective is to assess safety and tolerability of implantation of the Investigational Medicinal Product (ISTEM-01) in patients with retinitis pigmentosa.

Secondary objectives are:

- To evaluate the placement and position of the patch
- To assess preliminary efficacy based on:
- Evaluation of visual function
- Eye fundus
- Evaluation of photoreceptor survival

Assessment of visual function by Diagnosys-Full-field stimulus threshold (D-FST) is the only exploratory objective.

Official Title
A Phase 1/2, Open-label, Safety, Tolerability and Preliminary Efficacy Study of Implantation Into One Eye of hESC-derived RPE in Patients With Retinitis Pigmentosa Due to Monogenic Mutation

Conditions
Retinitis Pigmentosa

Study Type
Interventional

Enrollment (Actual)
7

Intervention/Treatment
Biological: Human Embryonic Stem Cell-Derived Retinal Pigment Epithelium (RPE)

Phase
Phase 1 Phase 2

Study Start (Actual)
2019-08-19

Primary Completion (Estimated)
2023-08-15

Study Completion (Estimated)
2026-12-15

France
Paris, France, 75012
Centre Hospitalier National d'Ophtalmologie (CHNO) des Quinze-Vingts

Eligibility Criteria
Ages Eligible For Study
18–65 years (adult, older adult)

Accepts Healthy Volunteer
No

Sexes Eligible for Study
All

Inclusion Criteria
- Documented diagnosis of retinitis pigmentosa based on a genetic test confirming the presence of a monogenic mutation that affects a gene involved in the visual signaling process specifically at the level of RPEs, namely RPE65 or LRAT, or MerTK
- 18 years old ≤Age ≤65 years old

For Patient of the First Cohort
- Visual acuity ≤20/200 in the best eye (legally blind)

For Patient of the Second Cohort

- 20/63 > Visual acuity >20/200 in the worst eye and
- Visible photoreceptor outer nuclear layer (ONL) on a spectral domain optical coherence tomography (OCT) scan

For the Two Cohorts

- Negative serum pregnancy test in women of childbearing potential (a woman who is 2 years post-menopausal confirmed by a physician or surgically sterile is not considered to be of childbearing potential)
- Female patients of childbearing potential (if sexually active), committed to use two methods of contraception starting from the enrollment, during mycophenolate mofetil (MMF) treatment and for 6 weeks after the last dose of MMF
- Sexually active men (including vasectomized men) committed to use condoms during from the first day of MMF treatment and for at least 90 days after cessation of treatment
- Signed informed consents by the patient or legal guardian(s). For patients unable to give consent, authorization to participate to the study will be collected close to their legally authorized representative
- Affiliated to or a beneficiary of a health care system

Exclusion Criteria

- Patient unable or unwilling to comply with the protocol requirements
- History of allergy or sensitivity to one of the products used during the study
- Patients with known serious allergies to the fluorescein
- Patients with a contraindication to general anesthesia
- Prior treatment with a gene or cell therapy product
- Patients with chronic hepatitis B or C, i.e., positive hepatitis B surface antigen or hepatitis C RNA viral load positive
- Patients infected with human immunodeficiency virus (HIV)
- Anti-HLA antibodies positive at screening
- Pregnancy or breastfeeding
- Presence of any ocular disease or ocular media opacity which in the opinion of the investigator precludes accurate evaluation
- Participation in another drug or device clinical study within last 6 months prior to baseline
- Patients known to be affected by pathologies for which the symptoms or associated treatments can alter the visual function and/or affect the retina
- Systemic corticosteroid therapy or other immunosuppressive/immunomodulating or anti-retroviral drugs within 2 months prior to baseline
- Patients with a contraindication to immunosuppressive/immunomodulating therapy (MMF) such as severe chronic renal impairment, severe digestive system disease, Lesch-Nyhan disease, Kelley-Seegmiller syndrome.
- Acute illness or infection within 4 weeks of the anticipated administration of study medication which may interfere with study assessments and immunosuppressive/immunomodulating therapy

- Any other condition or history that, in the opinion of the Investigator, may compromise the safety or compliance of the patient or would preclude the patient from successful completion of the study

Primary purpose: treatment
Allocation: N/A
Interventional model: sequential assignment
Interventional model description: open-label study
Masking: none (open label)

Arms and interventions

Participant group/arm	Intervention/treatment
Experimental: Implantation of a therapeutical patch All the patients will receive a single central subretinal implantation in one eye of a monolayer of human embryonic stem cell-derived retinal pigmented epithelium (hESC-derived RPE). The implanted eye will be the one with the worst visual acuity	Biological: Human embryonic stem cell-derived retinal pigment epithelium (RPE) • Implantation into one eye of hESC-derived RPE (human embryonic stem cell-derived retinal pigment epithelium (RPE))

Primary outcome measures

Outcome measure	Measure description	Time frame
Safety assessment	Safety and tolerability measured by the incidence of adverse events or serious adverse events throughout the study	From baseline until week 56

Secondary outcome measures

Outcome measure	Measure description	Time frame
Position of therapeutic patch	Position of the therapeutic patch by serial images assessment	From baseline until week 56
Placement of the therapeutic patch	Placement of the therapeutic patch by serial spectral domain ocular coherence tomography (OCT) scan at baseline and by study visit	From baseline until week 56
Change in leakage or perfusion	Change in leakage or perfusion in normal fundal vasculature and presence of abnormal vasculature by fundus fluorescein angiography	At baseline and weeks 24, 48, and 56
Change in thickness of RPE layer	Change in thickness of RPE layer by B-mode orbital ultrasound	At weeks 4, 8, 16, 24, 36, 48 and 56

Sponsor

Centre d'Etude des Cellules Souches

Collaborators

No information provided

Investigators

- Principal Investigator: Stéphane BERTIN, MD, Centre des 15–20

General Publications
No publications available.

Spain

Active, Not Recruiting

Clinical Trial to Evaluate Safety and Efficacy of Cell Therapy in Patients with Cicatricial Conjunctivitis (CELOPHIN)

ClinicalTrials.gov ID NCT05520086

Sponsor Instituto de Investigación Sanitaria de la Fundación Jiménez Díaz
Information provided by Instituto de Investigación Sanitaria de la Fundación Jiménez Díaz (Responsible Party)
Last Update Posted 2023-07-06

Brief Summary
A phase IIa, open-label, non-controlled clinical trial to assess the feasibility and safety of allogeneic adipose-derived mesenchymal stem cells (ASC) in the treatment of cicatricial conjunctivitis associated with Lyell's syndrome, Stevens-Johnson's syndrome, and mucous membrane pemphigoid with ocular involvement

Detailed Description
A multicenter, nationwide clinical trial will be performed. 20 patients fulfilling eligibility criteria will be included in two different dose level cohorts: the first 10 patients included will receive a single cell application and the remainder 10 will be administered 2 cell applications separated by 15 days.

The investigational drug consists of locally administered, expanded, allogeneic adipose-derived adult mesenchymal stem cells (ASCs), at a dose of 5 million cells per mL (0.5 mL to be infused per quadrant)

Study chronogram: recruitment is estimated to take 8 months, and follow-up period will be 12 months.

Conditions
Chronic conjunctivitis
Stevens-Johnson Syndrome
Lyell syndrome
Pemphigoid

Study Type
Interventional

Enrollment (Actual)
20

Intervention/Treatment
Drug: single dose
Drug: double dose

Phase
Phase 1 Phase 2

Other Study ID Numbers
Celophin

Study Start (Actual)
2022-11-11

Primary Completion (Estimated)
2024-02

Study Completion (Estimated)
2024-02

Spain
Córdoba, Spain, 14012
Hospital La Arruzafa
Madrid, Spain, 28007
Hospital General Universitario Gregorio Marañon
Madrid, Spain, 28040
Hospital Universitario Fundación Jiménez Diaz
Madrid, Spain, 28046
Hospital Universitario La Paz
Valladolid, Spain, 47003
Hospital Clínico Universitario de Valladolid

Asturias Locations
Oviedo, Asturias, Spain, 33012
Instituto Oftalmológico Fernández–Vega

Bizkaia Locations
Barakaldo, Bizkaia, Spain, 48903
Hospital Universitario de Cruces

Ciudad Real Locations
Alcázar De San Juan, Ciudad Real, Spain, 13600
Hospital General La Mancha Centro

Eligibility Criteria
Description

Inclusion Criteria
- Men and women over 18 years of age.
- Diagnosis of ocular pemphigoid in Foster stages I–IIcIIIb (2) or diagnosis of recurrent chronic or episodic inflammation accompanied by cicatricial conjunc-

tivitis of the mucous membranes with ocular involvement after the acute phase of Stevens-Johnson syndrome or Lyell syndrome with insufficient disease control or inacceptable toxicity or impossibility to administer usual care treatments (according to physician or patient criteria)
- In the case of women of childbearing age, who are willing to use an effective contraceptive method during the period of participation in the study
- Consent to participate and signature of the informed consent

Exclusion Criteria
- Signs of active infection on the ocular surface
- History of neoplasms in the last 5 years except for epithelial basal or squamous cell carcinoma
- Allergy to local anesthetics
- Patients who have participated in another clinical trial with medication during the 90 days prior to signing the IC
- Medical or psychiatric illness of any kind that, in the opinion of the investigator, may be a reason for exclusion from the study
- Congenital or acquired immunodeficiencies
- Major surgery or serious trauma of the subject in the semester prior to signing the IC
- Pregnant or lactating women
- Impossibility or refusal to carry out the follow-up required in the study by the patient

Ages Eligible for Study
18 years and older (adult, older adult)

Sexes Eligible for Study
All

Accepts Healthy Volunteers
No

Design Details
Primary purpose: treatment
Allocation: Non-randomized
Interventional model: sequential assignment
Interventional model description: 2 consecutive cohorts will be included in order of recruitment. the first 10 patients will receive 1 single dose and the second 10 patients will receive 2 doses separated by 15 days

Masking: None (open label)

Arms and interventions

Participant group/arm	Intervention/treatment
Experimental: Single dose allogenic adipose-derived mesenchymal stem cells Allogenic adipose-derived mesenchymal stem cells (MSC) Dose: 12.5 million cells	Drug: Single dose • Single dose: Administration of 12.5 million allogeneic mesenchymal stem cells derived from adipose tissue by subconjunctival injection and topical conjunctival injection
Experimental: Double-dose allogenic adipose-derived mesenchymal stem cells Allogenic adipose-derived mesenchymal stem cells (MSC) Dose: 12.5 million cells at day 0 and 12.5 million cells at day 14	Drug: Double dose • Repeated dose: administration of 2 doses, separated by 14 days, of 12.5 million allogeneic mesenchymal stem cells derived from adipose tissue by subconjunctival injection and topical conjunctival injection

Primary outcome measures

Outcome measure	Measure description	Time frame
Percentage of complications	Percentage of complications related to anesthesia, administration of investigational medicinal product and postoperative period	Through study completion, an average of 3 years

Secondary outcome measures

Outcome measure	Measure description	Time frame
Signs improvement	Changes from baseline in scarring conjunctivitis rating scale	At 7 days (and 21 days in the 2-dose group), 4 weeks, 12 weeks, 24 weeks, and 52 weeks
Symptom improvement	Changes from baseline in Ocular Surface Disease Index (OSDI) questionnaire (0–48). The lower the rate obtained in the score, better the outcome	At 7 days (and 21 days in the 2-dose group), 4 weeks, 12 weeks, 24 weeks, and 52 weeks
Visual acuity improvement	Changes from baseline in early treatment diabetic retinopathy study (ETDRS) (0–1.2) The higher the rate obtained in the score, better the outcome	At 7 days (and 21 days in the 2-dose group), 4 weeks, 12 weeks, 24 weeks, and 52 weeks
Changes in quality of life	Changes from baseline using the specific National Eye Institute-Visual Function Questionnaire 25 (NEI-VFQ-25) (0–100). The value of 0 is the lowest and 100 the best possible	At 7 days (and 21 days in the 2-dose group), 4 weeks, 12 weeks, 24 weeks, and 52 weeks
Evolution of the conjunctival flora	variation in the number and type of microorganisms in the conjunctival exudates, performing cultures and PCR herpes virus	At 7 days (and 21 days in the 2-dose group), 4 weeks, 12 weeks, 24 weeks, and 52 weeks

Sponsor

Instituto de Investigación Sanitaria de la Fundación Jiménez Díaz

Collaborators

- Effice Servicios Para la Investigacion S.L.

Investigators

- Principal Investigator: Nicolás Alejandre Alba, Hospital Fundación Jiménez Diaz

General Publications

No publications available.

Chapter 3
Completed Ocular Stem Cell Studies

Retina/Optic Nerve

United States

Safety of Repeat Intravitreal Injection of Human Retinal Progenitor Cells (jCell) in Adult Subjects With Retinitis Pigmentosa

ClinicalTrials.gov ID NCT04604899

Sponsor: jCyte, Inc.
Information provided by jCyte, Inc. (Responsible Party)
Last Update Posted 2022-08-10

Brief Summary
The primary objective of the study is to assess the safety of repeat injection of human retinal progenitor cells (jCell) in adult subjects with RP that have previously been treated with jCell.

Detailed Description
This is a prospective, multicenter, single-arm, Phase 2 study of human retinal progenitor cells (jCell) for the treatment of retinitis pigmentosa (RP). The study will include only subjects previously treated with jCell.

To assess reinjection of a previously treated eye, subjects who have previously been treated with jCell and desire a second treatment in the same eye will be enrolled. Subjects must have completed at least 12 months of follow-up since the prior injection of jCell. Subjects who have had both eyes previously treated with jCell will only have one eye retreated; the eye to be retreated will preferably be the better seeing eye, but exceptions may be made by the study investigator, taking into consideration BCVA, prior response to treatment, and any other medical conditions

J. N. Weiss, *Ocular Stem Cell Studies*,
https://doi.org/10.1007/978-3-031-58557-9_3

that may indicate which eye is the best candidate for retreatment. Subjects will be followed for 12 months for safety and efficacy.

Official Title
A Phase 2 Study of the Safety of Repeat Intravitreal Injection of Human Retinal Progenitor Cells (jCell) in Adult Subjects With Retinitis Pigmentosa (RP)

Conditions
Retinitis Pigmentosa

Study Type
Interventional

Enrollment (Actual)
30

Intervention/Treatment
Biological: human retinal progenitor cells

Phase
Phase 2

Other Study ID Numbers
JC02-2

Study Start (Actual)
2020-12-01

Primary Completion (Actual)
2022-03-22

Study Completion (Actual)
2022-03-22

United States

California Locations

Irvine, California, United States, 92697
Gavin Herbert Eye Inst, Univ Cal Irvine

Los Angeles, California, United States, 90074
Retina-Vitreous Associates Medical Group

Massachusetts Locations

Boston, Massachusetts, United States, 02114
Ophthalmic Consultants of Boston

Eligibility Criteria
Ages Eligible For Study
18 years and older (Adult, Older Adult)

Accepts Healthy Volunteer
No

Sexes Eligible for Study
All

Description

Inclusion Criteria

- Willing to give written informed consent, able to make the required study visits, and follow study protocol instructions.
- Completed the 12 months of follow-up in the subject's most recent jCell study and did not withdraw from the study for any reason.
- Adequate organ function:
  - blood counts (hematocrit, Hgb, WBC, platelets and differential) within normal range, or if outside of normal range, not clinically significant as judged by the investigator
  - liver function: alanine transaminase [ALT] and aspartate transaminase [AST] ≤2 times the upper limit of the normal range
  - total bilirubin ≤1.5 times the upper limit of the normal range
  - renal function: serum creatinine ≤1.25 times the upper limit of the normal range
- A female patient of childbearing potential (not surgically sterilized and less than one year postmenopausal) must have a negative pregnancy test (urine human chorionic gonadotropin) at entry (prior to injection) and must have used medically accepted contraception for at least one month prior to treatment. Women of childbearing potential and men must be advised to use a medically accepted method of contraception for at least 12 months following treatment.

Exclusion Criteria

- Malignancy, end-stage major organ disease (heart failure, significant arrhythmias, stroke or transient ischemic attacks, diabetes, immunosuppressive or autoimmune state, major psychiatric disorder, epilepsy, thyroid disease, COPD, renal failure, or any chronic systemic disease requiring continuous treatment with systemic steroids, anticoagulants, or immunosuppressive agents).
- History of eye disease other than RP that impairs visual function, including retinal vascular disease, elevated intraocular pressure/glaucoma, severe posterior uveitis, clinically significant macular edema, media opacity precluding visual exam, amblyopia and/or longstanding constant strabismus, as well as patients who require other intravitreal therapies.
- Allergy to penicillin or streptomycin.
- Adverse reaction to DMSO.
- Unable or unwilling to undergo pupil dilation, topical anesthesia, or any protocol-required procedure.
- Women who are nursing or who are planning to nurse during the 12 months that would follow study treatment.
- Any circumstance that in the opinion of the investigator, would interfere with participation in, or compliance with the study protocol.

- Treatment with corticosteroids (systemic, periocular, or intravitreal) or any other non-approved, experimental, investigational or neuroprotectant therapy (systemic, topical, intravitreal) in either eye within 90 days of planned second injection.
- Cataract surgery within three months prior to treatment or anticipated to need cataract surgery within a year of treatment.

Design Details

Primary Purpose: Treatment
Allocation: N/A
Interventional Model: Single Group Assignment
Masking: None (Open Label)

Arms and interventions

Participant group/arm	Intervention/treatment
Experimental: Retreated subjects Subjects receiving human retinal progenitor cells (jCell) who have previously received jCell is a jCyte study	Biological: human retinal progenitor cells • Single intravitreal injection of 6.0 million human retinal progenitor cells (hRPC) • Other Names: – jCell

Primary outcome measures

Outcome measure	Measure description	Time frame
Safety of intravitreal injection of hRPC	Assessed by proportion of subjects with treatment emergent adverse events	12 months

Secondary outcome measures

Outcome measure	Measure description	Time frame
Best-corrected visual acuity	Assessed by E-ETDRS	12 months
Visual fields	The Octopus 900 will be used for kinetic visual field testing using a specified target of V4e for more severe subjects and a target of III4e and V4e for better seeing subjects	12 months
Contrast sensitivity	The Beethoven System will be used to capture the peak mean contrast sensitivity threshold value at any given spatial frequency	12 months
Mobility	Maze testing	12 months

Sponsor

jCyte, Inc.

Collaborators

1. California Institute for Regenerative Medicine (CIRM)

Investigators

1. Principal Investigator: Mitul Mehta, MD, University of California, Irvine/Gavin Herbert Eye Institute
2. Principal Investigator: David Liao, MD, Retina-Vitreous Associates Medical Group, Los Angeles CA
3. Principal Investigator: Anthony Jospeh, MD, Ophthalmic Consultants of Boston

General Publications
No publications available

Completed

Safety and Efficacy of Intravitreal Injection of Human Retinal Progenitor Cells in Adults With Retinitis Pigmentosa

ClinicalTrials.gov ID NCT03073733

Sponsor: jCyte, Inc.
Information provided by jCyte, Inc. (Responsible Party)
Last Update Posted 2022-04-04

Brief Summary
This study evaluates the changes in visual function at 12 months following a single injection of human retinal progenitor cells compared to sham-treated controls in a cohort of adult subjects with RP.

Detailed Description
There is no effective treatment for RP; once photoreceptors are lost, they do not regenerate. The rate of deterioration of vision varies from person to person, with most people with RP legally blind by age 40. Preclinical studies demonstrated that transplantation of retinal progenitor cells into the eye can result in both photoreceptor replacement and significant slowing of host photoreceptor loss. Thus, the primary goal of this therapy is to preserve, and potentially improve, vision by intervening in the disease at a time when dystrophic host photoreceptors can be protected and reactivated. Based on the demonstration of acceptable safety and tolerability in a phase 1/2a study, this phase 2b study is designed as a controlled comparison of the changes in visual function and functional vision in subjects who receive a single jCell injection in comparison to a comparable sham-treated control group of subjects with RP.

Official Title
A Prospective, Multicenter, Randomized, Study of the Safety and Efficacy of Intravitreal Injection of Human Retinal Progenitor Cells (jCell) in Adult Subjects With Retinitis Pigmentosa (RP)

Conditions
Retinitis Pigmentosa

Study Type
Interventional

Enrollment (Actual)
84

Intervention/Treatment
Biological: human retinal progenitor cells
Other: Mock injection

Phase
Phase 2

Other Study ID Numbers
JC-02

Study Start (Actual)
2017-03-01

Primary Completion (Actual)
2020-11-13

Study Completion (Actual)
2020-11-13

Contacts and Locations
United States
California Locations

Irvine, California, United States, 92697
Gavin Herbert Eye Inst, Univ Cal Irvine

Los Angeles, California, United States, 90,074
Retina-Vitreous Associates Medical Group

Massachusetts Locations

Boston, Massachusetts, United States, 02114
Ophthalmic Consultants of Boston

Eligibility Criteria
Ages Eligible For Study
18 years and older (Adult, Older Adult)

Accepts Healthy Volunteer
No

Sexes Eligible For Study
All

Description

Inclusion Criteria
Clinical diagnosis of RP confirmed by ERG and willing to consent to mutation typing, if not already done Best-corrected visual acuity (BCVA) 20/80 or worse and no worse than 20/800 Adequate organ function and negative infectious disease screen Female of childbearing potential must have negative pregnancy test and be willing to use medically accepted methods of contraception throughout the study.

Exclusion Criteria

Eye disease other than RP that impairs visual function Pseudo-RP, cancer-associated retinopathies, History of malignancy or other end-stage organ disease, or any chronic disease requiring continuous treatment with system steroids, anticoagulants, or immunosuppressive agents. Known allergy to penicillin or streptomycin treatment with corticosteroids or any investigational or neuroprotectant therapy within 90 days of enrollment Cataract surgery within 3 months prior to enrollment.

Study Plan

Design Details

Primary Purpose: Treatment
Allocation: Randomized
Interventional Model: Parallel Assignment
Masking: Double

Arms and interventions

Participant group/arm	Intervention/treatment
Experimental: Test (jCell injection) dose level 1 single intravitreal injection of 3.0 × 10e6 human retinal progenitor cells into the eye with the poorest visual acuity or, if vision is comparable in both eyes, the nondominant eye	Biological: human retinal progenitor cells • Live suspension of 3.0 or 6.0 × 10e6 human retinal progenitor cells (hRPC) suspended in clinical-grade medium injected intravitreally under local anesthesia • Other Names: – jCell
Other: Sham-treated Control a mock injection will be performed on the eye with the poorest vision in each Control subject (designated as the "study eye")	Other: Mock injection • Pressing the hub of a syringe with no needle against the eye to mimic intravitreal injection
Experimental: test (jCell injection) dose level 2 single intravitreal injection of 6.0 × 10e6 human retinal progenitor cells into the eye with the poorest visual acuity or, if vision is comparable in both eyes, the nondominant eye	Biological: human retinal progenitor cells • Live suspension of 3.0 or 6.0 × 10e6 human retinal progenitor cells (hRPC) suspended in clinical-grade medium injected intravitreally under local anesthesia • Other Names: – jCell

Primary outcome measures

Outcome measure	Measure description	Time frame
Best-corrected visual acuity (BCVA)	Mean change in BCVA in study eye from baseline to month 12 as assessed by E-ETDRS in ITT population. A letter score is used to compare change over time, with a higher number of letters representing better visual function, and a lower number of letters representing worse visual function. For example, 85 letters is equivalent to 20/20 visual acuity and 5 letters is equivalent to 20/800 visual acuity. A change value is derived for each subject by taking the letter score at 12 months and subtracting the letter score at baseline. A mean of all change values is then calculated for each arm	12 months

Secondary outcome measures

Outcome measure	Measure description	Time frame
Contrast sensitivity (CS) at 1.0 CPD	Assessment of the ability to detect changes in shades of grey, as measured with a vertical striped pattern that varies in width (cycles per degree or CPD); CS thresholds are created by taking the mean of multiple trials at each pattern width. A CS curve is created using a minimum of three pattern widths, one at the subject's peak or highest sensitivity, and then one larger and smaller pattern width on either side of the peak. For example, if a subject's CS peak appears to be at 2.0 CPD, then additional testing occurs at 1.0 and 4.0 CPD. RP patients have suppressed CS curves and the most common pattern widths are at 0.5, 1.0, 2.0, and 4.0 CPD. Severely impaired subjects (e.g., BCVA <20/400) usually have flat curves with low values (e.g., 1.28), while in mildly impaired RP subjects the values can be higher (e.g., 7.12), but are rarely near normal. These data represent mean change in CS from baseline to 12 months, with greater values representing greater improvement in visual function	12 months
Kinetic visual field (KVF)	Mean change in total area (degrees squared) of all islands of vision from baseline to 12 months	12 months
Low luminance mobility test (LLMT)	The LLMT identifies the performance of RP patients as they walk along an indoor pathway of arrows and obstacles at varying lighting levels. The Critical Illumination Level (CIL) is the light level below which the patient has a markedly slower pace and more errors than all light levels above (brighter than) that point. The LLMT uses light levels that go from very dim (0.12 lux) to a bright indoor room (500 lux), with evenly spaced increments that increase light by doubling the brightness of the room from the prior level. These evenly spaced light levels have been converted to a scale score to enable easier calculation of change scores. The dimmest light level of 0 lux (completely dark room) corresponds to a scale score of 13, whereas the brightest light level of 500 lux corresponds to a scale score of 0. A positive scale score change from baseline to 12 months represents improvement in low light vision, whereas a negative scale score change represents a decline in low light vision	12 months
Low vision functional questionnaire (visual ability)	The VA LV VFQ-48 (VFQ) is used to capture changes in patients' self-reporting of their difficulty with reading, mobility, and performing other daily living activities affected by visual impairment. There are four scales on the VFQ including Visual Information, Reading, Visual Motor, and Mobility. A fifth value, Visual Ability, is an aggregate score of the four scales, measured in units called logits, and is calculated for each person based on item weighting using Rasch analysis. Visual Ability is used broadly to represent changes in subject-reported outcomes from visit to visit. Higher positive values on the Visual Ability score represent better function and less impairment, whereas lower or negatives scores represent worse function or more impairment. Change in the Visual Ability scale on the VFQ is calculated by taking a subject's score at 12 months and subtracting from it the baseline score	12 months

Safety of intravitreal injection of retinal progenitor cells (RPC)	Assessed by treatment emergent adverse events, immunogenicity, and safety visual assessments	12 months

Sponsor
jCyte, Inc.

Collaborators
- California Institute for Regenerative Medicine (CIRM)

Investigators
- Principal Investigator: Baruch Kuppermann, MD, UCI

General Publications
No publications available

Results
Best-Corrected Visual Acuity (BCVA)
Type: Primary|Time Frame: 12 months

Description	Mean change in BCVA in study eye from baseline to month 12 as assessed by E-ETDRS in ITT population. A letter score is used to compare change over time, with a higher number of letters representing better visual function, and a lower number of letters representing worse visual function. For example, 85 letters is equivalent to 20/20 visual acuity and five letters is equivalent to 20/800 visual acuity. A change value is derived for each subject by taking the letter score at 12 months and subtracting the letter score at baseline. A mean of all change values is then calculated for each arm
Time frame	12 months
Analysis population description	ITT: All randomized subjects who provide any post-randomization data. One subject received a 4.0 × 10e6 dose prior to a protocol amendment that excluded this dose level; this subject is not included in these analyses. Also, of the 83 remaining subjects to be enrolled and randomized, 80 completed the study

Arm/group title	Sham-treated control	Test (jCell Injection) dose level 1	Test (jCell Injection) dose level 2
Arm/group description	Mock injection: pressing the hub of a syringe with no needle against the study eye to mimic intravitreal injection Show more	Single intravitreal injection of 3.0 × 10e6 retinal progenitor cells into the study eye	Single intravitreal injection of 6.0 × 10e6 retinal progenitor cells into the study eye
Overall number of participants analyzed	28	26	26
Mean (standard deviation)|Unit of measure: letters correct	3.9 (9.92)	1.3 (14.14)	2.6 (21.53)

Completed

Safety of a Single, Intravitreal Injection of Human Retinal Progenitor Cells (jCell) in Retinitis Pigmentosa

ClinicalTrials.gov ID NCT02320812

Sponsor: jCyte, Inc.
Information provided by jCyte, Inc. (Responsible Party)
Last Update Posted 2019-03-05

Brief Summary
This study evaluates the safety and potential activity of a single dose of live human retinal progenitor cells (jCell) administered to adults with retinitis pigmentosa. Four different dose levels of cells will be assessed in each of two groups of patients.

Detailed Description
The primary purpose of this study is to test the safety and tolerability of the administration of a single dose of jCell to adults with retinitis pigmentosa.

The goal of jCell therapy is to preserve vision by intervening in the disease at a time when host photoreceptors can be protected and potentially reactivated.

Human allogeneic retinal progenitor cells will be injected into adults with advanced RP to see if the procedure is safe, if the cells survive, and whether they have any impact on the visual status of the patients.

Official Title
A Prospective, Multicenter, Open-Label, Single-Arm Study of the Safety and Tolerability of a Single, Intravitreal Injection of Human Retinal Progenitor Cells (jCell) in Adult Subjects With Retinitis Pigmentosa (RP)

Conditions
Retinitis Pigmentosa (RP)

Study Type
Interventional

Enrollment (Actual)
28

Intervention/Treatment
Biological: human retinal progenitor cells

Phase
Phase 1 Phase 2

Other Study ID Numbers
JC-01

Study Start
2015-06

Primary Completion (Actual)
2017-07-19

Study Completion (Actual)
2017-07-19

United States

California Locations

Irvine, California, United States
The Gavin Herbert Eye Institute, Univ California Irvine

Los Angeles, California, United States, 90,074
Retina-Vitreous Associates Medical Group

Eligibility Criteria
Ages Eligible For Study
18 years and older (Adult, Older Adult)

Accepts Healthy Volunteer
No

Sexes Eligible for Study
All

Description

Inclusion Criteria
- Clinical diagnosis of RP confirmed by electroretinogram (ERG) and willing to consent to mutation typing, if not already done
- Best-corrected visual acuity (BCVA) 20/63 or worse and no worse than hand motions (HM)
- Adequate organ function and negative infectious disease screen
- Female of childbearing potential must have negative pregnancy test and be willing to use medically accepted methods of contraception throughout the study

Exclusion Criteria
- Eye disease other than RP that impairs visual function
- Pseudo-RP, cancer-associated retinopathies
- History of malignancy or other end-stage organ disease, or any chronic disease requiring continuous treatment with system steroids, anticoagulants, or immunosuppressive agents
- Known allergy to penicillin or streptomycin

Design Details
Primary Purpose: Treatment
Allocation: N/A
Interventional Model: Single Group Assignment
Masking: None (Open Label)

Arms and interventions

Participant group/arm	Intervention/treatment
Experimental: Treated subjects human retinal progenitor cells	Biological: human retinal progenitor cells • Single intravitreal injection of 0.5–3.0 million human retinal progenitor cells (hRPC) • Other Names: – jCell

Primary outcome measures

Outcome measure	Measure description	Time frame
Number of subjects with adverse events as a measure of safety and tolerability	Proportion of subjects with treatment emergent adverse events (TEAE), related TEAE, and severe TEAE	12 months

Secondary outcome measures

Outcome measure	Measure description	Time frame
Change in mean best- corrected visual acuity (BCVA)	Change in mean best-corrected visual acuity in test eye versus untreated eye; BCVA is measured using an eye chart and is reported as the number of letters read correctly (ranging from 0 to 100 letters) in the study eye. The lower the number of letters read correctly on the eye chart, the worse the vision (or visual acuity). A positive change from baseline indicates an improvement and a negative change from baseline indicates a worsening. Change between baseline and month 12 is reported.	12 months

Sponsor

jCyte, Inc.

Collaborators

- California Institute for Regenerative Medicine (CIRM)

Investigators

- Principal Investigator: Baruch Kuppermann, MD, Gavin Herbert Eye Institute, UCI, Irvine, CA

General Publications

No publications available

Results

Publications About Study Results

No publications available

Results

1. Number of Subjects with Adverse Events as a Measure of Safety and Tolerability

Type: Primary | Time Frame: 12 months

Description	Proportion of subjects with treatment emergent adverse events (TEAE), related TEAE, and severe TEAE
Time frame	12 months
Analysis population description	All treated subjects

Arm/group title	Treated subjects
Arm/group description	Human retinal progenitor cells Human retinal progenitor cells: single intravitreal injection of 0.5–3.0 million human retinal progenitor cells (hRPC)
Overall number of participants analyzed	28
Subjects with TEAE [a]Measure type: count of participants | Unit of measure: participants	2589.3%
Subjects with related TEAE[a]	2175.0%
Subjects with grade 3 TEAE[a]	13.6%

[a]Measure Type: Count of Participants | Unit of Measure: Participants

2. Change in Mean Best-Corrected Visual Acuity (BCVA)
   Type: Secondary|Time Frame: 12 months

Description	Change in mean best-corrected visual acuity in test eye versus untreated eye; BCVA is measured using an eye chart and is reported as the number of letters read correctly (ranging from 0 to 100 letters) in the study eye. The lower the number of letters read correctly on the eye chart, the worse the vision (or visual acuity). A positive change from baseline indicates an improvement and a negative change from baseline indicates a worsening. Change between baseline and month 12 is reported
Time frame	12 months
Analysis population description	All treated subjects

Arm/group title	0.5 Million cells cohort	1.0 Million cells cohort	2.0 Million cells cohort	3.0 Million cells cohort
Arm/group description	Subjects who received 0.5 million human retinal progenitor cells (hRPC)	Subjects who received 1.0 million hRPC	Subjects who received 2.0 million hRPC	Subjects who received 3.0 million hRPC
Overall number of participants analyzed	8	8	6	6
Mean (full range) | Unit of measure: number of letters read correctly	1.4 (−3 to 11)	1.0 (−3 to 11)	4.8 (0 to 8)	9.0 (0 to 12)

Adverse Events

Time Frame

one year following single intravitreal injection of hRPC

Adverse event reporting description

Arm/group title	Treated subjects	
Arm/group description	Human retinal progenitor cells Human retinal progenitor cells: single intravitreal injection of 0.5–3.0 million human retinal progenitor cells (hRPC)	
All-cause mortality		
Arm/group title	Treated Subjects	
	Affected/at Risk (%)	
Total	0/28 (0.00%)	
Serious adverse events		
Arm/group title	Treated Subjects	
	Affected/at Risk (%)	# Events
Total	1/28 (3.57%)	
Musculoskeletal and connective tissue disorders		
Migratory pain[a,b,c]	1/28 (3.57%)	1
[a] Indicates events were collected by systematic assessment [b] Term from vocabulary, MedDRA [c] Grade 2		
5% Frequency Threshold for Reporting Other Adverse Events		
Arm/group title	Treated Subjects	
	Affected/at Risk (%)	# Events
Total	20/28 (71.43%)	
Eye disorders		
Anterior chamber flare[a,b]	5/28 (17.86%)	5
Conjunctival hemorrhage[a,b]	10/28 (35.71%)	10
Eye pain[a,b]	5/28 (17.86%)	5

[a]Indicates events were collected by systematic assessment
[b]Term from vocabulary, MedDRA
[c]Grade 2

Completed

Hematopoietic Stem Cell Transplant in Devic's Disease

ClinicalTrials.gov ID NCT00787722

Sponsor: Northwestern University
Information provided by Richard Burt, MD, Northwestern University (Responsible Party)
Last Update Posted 2020-02-28

Brief Summary

This study is designed to examine whether treating Devic's disease patients with high-dose cyclophosphamide together with rabbit antithymocyte globulin (rATG)/ rituximab (drugs which reduce the function of the immune system), followed by return of previously collected patient's stem cells will result in improvement in Devic's disease. Stem cells are undeveloped cells that have the capacity to grow into mature blood cells, which normally circulate in the blood stream. The purpose of the intense chemotherapy is to destroy the cells in patient's immune system, which may be causing his/her disease. The purpose of the stem cell infusion is to produce a normal immune system that will no longer attack patient's body. The purpose of study is to examine the safety and efficacy of this treatment. The drugs used in this study treatment are drugs for commonly used for immune suppression.

Detailed Description

Neuromyelitis optica (NMO, Devic's disease) is an autoimmune, inflammatory, demyelinating central nervous system disorder in which a person's own immune system attacks the optic nerves and spinal cord and is characterized by concurrence of optic neuritis and transverse myelitis, typically associated with a lesion in the spinal cord extending over three or more vertebral segments. Although it is most commonly relapsing, it is distinct from multiple sclerosis in that it is more severe, tends to spare the brain, and is associated with a longitudinally extensive lesion on spinal cord MRI. Furthermore, NMO is associated with a highly specific serum autoantibody marker, NMO-immunoglobulin G (IgG), which targets the water channel aquaporin-4. The disease follows a relapsing course in more than 90% of patients.

At present, parenteral corticosteroids are widely employed as first-line treatment of optic neuritis and myelitis attacks, whereas therapeutic plasmapheresis is applied in the case of corticosteroids failure. Various strategies for the prevention of NMO relapses have been employed in small case series with modest activity. Immune-based therapies, in order to be effective, need to be started early in the disease course while Devic's disease is predominantly an immune-mediated and inflammatory disease. Since 50% of patients with NMO are confined to a wheelchair within 5 years of onset, new therapies are needed in this disease.

We now propose, as a phase I study, complete immune ablation and subsequent reconstitution with autologous stem cells. Based on the experience of the pilot studies, the current protocol will mobilize stem cells with granulocyte-colony stimulating factor (G-CSF) and cyclophosphamide and collect stem cells by apheresis. A subsequent bone marrow harvest will be performed only if needed to supplement the peripheral blood stem cells (PBSC). Based on the experience of autoimmune flares in patients receiving G-CSF alone for mobilization, patients will be mobilized with cyclophosphamide 2.0 g/m^2 and G-CSF 5–10 mcg/kg.

In order to avoid cumulative cardiac toxicity from cyclophosphamide and to allow culture of hematopoietic stem cell (HSC) product, three weeks must separate the administration of cyclophosphamide for mobilization and for conditioning.

Cyclophosphamide 50 mg/kg/day will be given IV over 2 h in 500 cc of normal saline. If actual weight is < ideal weight, cyclophosphamide will be given based on actual weight. If actual weight is > ideal weight, cyclophosphamide will be given as adjusted weight. Adjusted weight = ideal weight + 25% (actual weight minus ideal weight).

Hydration-guidelines, normal saline (NS) at 150–200 ml/h should be given 2 h before cyclophosphamide and continued until 24 h after the last cyclophosphamide dose. The rate of hydration will be aggressively adjusted. Twice daily weights will be obtained. Amount of fluid can be modified based on patient's fluid status.

r ATG 0.5 mg/kg given on day-5, then 1.0 mg/kg given on day −4, then 1.5 mg/kg given on days −3 through −1. rATG is infused over 10 h. Premedicate with Acetaminophen 650 mg po and Diphenhydramine 25 mg po/IV 30 min before the infusion.

Rituxan (Rituximab)—The dose of 500 mg of Rituximab will be diluted in 500 ml 0.9% NS and infused per standard Rituximab infusion guidelines, given on days −6 and on day +1. Following the guidelines, Rituximab will be started at 50 mg/h. If no reaction occurs, the dose will be increased by 50 mg/h every 30 min to a maximum of 400 mg/h G-CSF-guidelines, 5–10 mcg/kg/day will be started day +5 and continued until the absolute neutrophil counts reaches at least 1000/μl.

Official Title
Trial of High-Dose Immunosuppressive Therapy With Hematopoietic Stem Cell Support in Devic's Disease

Conditions
Devic's Disease

Study Type
Interventional

Enrollment (Actual)
13

Intervention/Treatment
Procedure: Hematopoietic Stem Cell Transplantation
Drug: Cyclophosphamide
Drug: G-CSF
Drug: rATG
Drug: Mesna
Drug: Rituximab
Drug: Methylprednisolone

Phase
Phase 1 Phase 2

Other Study ID Numbers
DIAD Devic's Disease Auto 2008

Study Start (Actual)
2009-10-10

Primary Completion (Actual)
2018-11

Study Completion (Actual)
2018-11

United States
Illinois Locations

Chicago, Illinois, United States, 60611
Northwestern University, Feinberg School of Medicine

Ages Eligible for Study
16–65 years (Child, Adult, Older Adult)

Accepts Healthy Volunteer
No

Sexes Eligible for Study
All

Description

Inclusion Criteria
- Age between 16–65, at the time of pretransplant evaluation
- An established diagnosis of Devic's disease (more than one acute attack)
- NMO- IgG aquaporin-4 autoantibody positive

Exclusion Criteria
- Paraplegia or quadriplegia and legal blindness (defined as visual acuity of 20/200 or less in the better eye with the best correction possible)
- Any illness that in the opinion of the investigators would jeopardize the ability of the patient to tolerate aggressive chemotherapy
- Prior history of malignancy except localized basal cell, squamous skin cancer, or carcinoma in situ of the cervix. Other malignancies for which the patient is judged to be cured, such as head and neck cancer, or breast cancer will be considered on an individual basis
- Positive pregnancy test
- Inability or unwillingness to pursue effective means of birth control. Effective birth control is defined as (1) refraining from all acts of vaginal intercourse (ABSTINENCE); (2) consistent use of birth control pills; (3) injectable birth control methods (Depo-provera, Norplant); (4) tubal sterilization or male partner who has undergone vasectomy; (5) placement of an intrauterine device (IUD); or (6) use, with every act of intercourse, of diaphragm with contraceptive jelly and/or condoms with contraceptive foam
- Failure to willingly accept or comprehend irreversible sterility as a side effect of therapy
- Forced expiratory volume at one (FEV1)/forced vital capacity (FVC) < 60% of predicted after bronchodilator therapy (if necessary)
- Diffusing capacity of lung for carbon monoxide (DLCO) < 50% of predicted

- Resting left ventricular ejection fraction (LVEF) < 50%
- Serum creatinine >2.0 mg/dl
- Known hypersensitivity to mouse, rabbit, or E. coli-derived proteins
- Presence of metallic objects implanted in the body that would preclude the ability of the patient to safely have MRI exams
- Bilirubin >2.0 mg/dl
- Platelet count <100,000/ul or absolute neutrophil count (ANC) < 1000/ul
- Psychiatric illness, mental deficiency, or cognitive dysfunction making compliance with treatment or informed consent impossible
- Active infection except asymptomatic bacteriuria
- Inability to give informed consent
- HIV positive
- Transaminases >3× of normal limits, liver cirrhosis

Design Details
Primary Purpose: Treatment
Allocation: N/A
Interventional Model: Single Group Assignment
Masking: None (Open Label)

Arms and interventions

Participant group/arm	Intervention/treatment
Experimental: Hematopoietic Stem Cell Transplantation Hematopoietic stem cell transplantation will be performed after conditioning regimen of cyclophosphamide, G-CSF, Mesna, rATG, rituximab, and methylprednisolone	Procedure: Hematopoietic Stem Cell Transplantation • Infusion of participant's own stem cells Drug: Cyclophosphamide • A medication used as chemotherapy and to suppress the immune system • Other Names: – Cytoxan – Neosar Drug: G-CSF • A glycoprotein that stimulates the bone marrow to produce granulocytes and stem cells and release them into the bloodstream • Other Names: – Neupogen – Filgrastim – Granix – Zarxio Drug: rATG • A rabbit polyclonal antibody to lymphocytes • Other Names: – Thymoglobulin – Anti-Thymocyte Globulin Drug: Mesna • A medication used in those taking cyclophosphamide or ifosfamide to decrease the risk of bleeding from the bladder • Other Names: – Mesnex Drug: Rituximab • Monoclonal antibody therapy used to treat certain autoimmune diseases and types of cancer • Other Names: – Rituxan Drug: Methylprednisolone • A corticosteroid medication used to suppress the immune system and decrease inflammation • Other Names: – Solu-Medrol – Depo-Medrol

Primary outcome measures

Outcome measure	Measure description	Time frame
Survival	Survival rate will be evaluated at 6 months, 1 year, 2 year, 3 year, 4 year, 5 year after the transplant	6 months, 1 year, 2 year, 3 year, 4 year, 5 year—after the transplant

Secondary outcome measures

Outcome measure	Measure description	Time frame
Quality of life (QOL) Short Form-36 (SF-36)	SF-36 is a self-administered quality of life exam. The evaluation of the results was done by attributing scores to each question, which were then transformed into a scale ranging from 0 to 100, where 0 corresponds to the worst quality of life and 100 to the best.	Pretransplant 12 months and 5 years
Post HSCT immune - modulating medication and relapse	Number of immune-modulating medication and relapse evaluated 5 year—after the transplant	Pretransplant and 6 months, 1 year, 2 year, 3 year, 4 year, and 5 year after transplant
Number of patients who require no device assistance for ambulation	No Device Assistance Needed for Ambulation evaluated at 6 months, 1 year, 2 year, 3 year, 4 year, 5 year—after the transplant	6 months, 1 year, 2 year, 3 year, 4 year, 5 year—after the transplant
Disability score: expanded disability status scale (EDSS)	Disability scores (disease improvement defined by at least a 1 point increase in the Expanded Disability Status Scale (EDSS) on consecutive evaluations at least three months apart The EDSS scale ranges from 0 to 10 in 0.5 unit increments that represent higher levels of disability	Pretransplant 6 month, 5 year
NMO-IgG Aquaporin-4 autoantibody titer	NMO-IgG aquaporin-4 autoantibody titer will be tested pretransplant and posttransplant	Pretransplant and 5 year Posttransplant

Sponsor

Northwestern University

Collaborators

No information provided

Investigators

- Principal Investigator: Richard Burt, MD, Northwestern University

General Publications

- Burt RK, Balabanov R, Han X, Burns C, Gastala J, Jovanovic B, Helenowski I, Jitprapaikulsan J, Fryer JP, Pittock SJ. Autologous nonmyeloablative hematopoietic stem cell transplantation for neuromyelitis optica. Neurology. 2019 Oct 29;93(18):e1732-e1741. https://doi.org/10.1212/WNL.0000000000008394. Epub 2019 Oct 2.

Results

1. Survival

   Type: Primary | Time Frame: 6 months, 1 year, 2 year, 3 year, 4 year, 5 year—after the transplant

Description	Survival rate will be evaluated at 6 months, 1 year, 2 year, 3 year, 4 year, 5 year after the transplant
Time frame	6 months, 1 year, 2 year, 3 year, 4 year, 5 year—after the transplant
Analysis population description	The reason that the number analyzed in one or more rows differs from the overall number analyzed is because one patient with coexistent SLE died of complications from active lupus 10 months after the transplant and another patient was only 3 years out from transplant

Arm/group title	Hematopoietic stem cell transplantation
Arm/group description	Hematopoietic stem cell transplantation will be performed after conditioning regimen of cyclophosphamide, G-CSF, Mesna, rATG, rituximab, and methylprednisolone Hematopoietic Stem Cell Transplantation: Infusion of participant's own stem cells Cyclophosphamide: A medication used as chemotherapy and to suppress the immune system G-CSF: A glycoprotein that stimulates the bone marrow to produce granulocytes and stem cells and release them into the bloodstream rATG: A rabbit polyclonal antibody to lymphocytes Mesna: A medication used in those taking cyclophosphamide or ifosfamide to decrease the risk of bleeding from the bladder Rituximab: Monoclonal antibody therapy used to treat certain autoimmune diseases and types of cancer Methylprednisolone: A corticosteroid medication used to suppress the immune system and decrease inflammation
Overall number of participants analyzed	13
6 months survival	
Number analyzed	13 participants
[a]Measure type: count of participants | Unit of measure: participants	13100.0%
1 year survival	
Number analyzed	12 participants
[a]	12100.0%
2 year survival	
Number analyzed	12 participants
[a]	12100.0%
3 year survival	
Number analyzed	11 participants
[a]	11100.0%
4 year survival	
Number analyzed	11 participants

[a]	11100.0%
5 year survival	
Number analyzed	11 participants
[a]	11100.0%

[a]Measure type: count of participants | unit of measure: participants

2. Quality Of Life (QOL) Short Form-36 (SF-36)
   Type: Secondary|Time Frame: pretransplant 12 months and 5 years

Description	SF-36 is a self-administered quality of life exam. The evaluation of the results was done by attributing scores to each question, which were then transformed into a scale ranging from 0 to 100, where 0 corresponds to the worst quality of life and 100 to the best
Time frame	Pretransplant 12 months and 5 years
Analysis population description	The reason that the number analyzed differs in one or more rows is because one patient did not complete the form and was excluded. Another patient was not 5 years out from transplant so there is no 5 year data on the patient

Arm/group title	Hematopoietic stem cell transplantation
Arm/group description	Hematopoietic stem cell transplantation will be performed after conditioning regimen of cyclophosphamide, G-CSF, Mesna, rATG, rituximab, and methylprednisolone Hematopoietic Stem Cell Transplantation: Infusion of participant's own stem cells Cyclophosphamide: A medication used as chemotherapy and to suppress the immune system G-CSF: A glycoprotein that stimulates the bone marrow to produce granulocytes and stem cells and release them into the bloodstream rATG: A rabbit polyclonal antibody to lymphocytes Mesna: A medication used in those taking cyclophosphamide or ifosfamide to decrease the risk of bleeding from the bladder Rituximab: Monoclonal antibody therapy used to treat certain autoimmune diseases and types of cancer Methylprednisolone: A corticosteroid medication used to suppress the immune system and decrease inflammation
Overall number of participants analyzed	11
Pretransplant	
Number analyzed	11 participants
[a]Median (standard deviation) | unit of measure: score on a scale	30.83 (34.22)
1 year posttransplant	
Number analyzed	11 participants
[a]	52.69 (55.08)
5 year posttransplant	
Number analyzed	10 participants
[a]	61.63 (62.09)

[a]Median (standard deviation) | unit of measure: score on a scale

3. Post Hsct Immune-Modulating Medication and Relapse
   Type: Secondary|Time Frame: Pre transplant and 6 months, 1 year, 2 year, 3 year, 4 year, and 5 year after transplant

Description	Number of immune-modulating medication and relapse evaluated 5 year—after the transplant
Time frame	Pretransplant and 6 months, 1 year, 2 year, 3 year, 4 year, and 5 year after transplant
Analysis population description	The reason that the number analyzed in one or more rows differs from the overall number analyzed is because one patient with coexistent SLE died of complications from active lupus 10 months after the transplant and another patient was only 3 years out from transplant

Arm/group title	Hematopoietic stem cell transplantation (HSCT)
Arm/group description	Hematopoietic stem cell transplantation will be performed after conditioning regimen of cyclophosphamide, G-CSF, Mesna, rATG, rituximab, and methylprednisolone Hematopoietic Stem Cell Transplantation: Infusion of participant's own stem cells Cyclophosphamide: A medication used as chemotherapy and to suppress the immune system G-CSF: A glycoprotein that stimulates the bone marrow to produce granulocytes and stem cells and release them into the bloodstream rATG: A rabbit polyclonal antibody to lymphocytes Mesna: A medication used in those taking cyclophosphamide or ifosfamide to decrease the risk of bleeding from the bladder Rituximab: Monoclonal antibody therapy used to treat certain autoimmune diseases and types of cancer Methylprednisolone: A corticosteroid medication used to suppress the immune system and decrease inflammation
Overall number of participants analyzed	12
Pre HSCT—Immunosuppression/ rclapsc ratc	
Number analyzed	12 participants
[a]Measure type: count of participants | unit of measure: participants	12100.0%
6 Mos Post HSCT immunosuppression/relapse rate	
Number analyzed	12 participants
[a]	18.3%
1 year post HSCT immunosuppression/relapse rate	
Number analyzed	11 participants
[a]	19.1%
2 year post HSCT immunosuppression/relapse rate	
Number analyzed	12 participants

a	325.0%
3 year post HSCT immunosuppression/relapse rate	
Number analyzed	11 participants
a	00.0%
4 year post HSCT immunosuppression/relapse rate	
Number analyzed	11 participants
a	19.1%
5 year post HSCT immunosuppression relapse rate	
Number analyzed	11 participants
a	19.1%

[a]Measure type: count of participants | unit of measure: participants

4. Number of Patients who Require no Device Assistance for Ambulation
   Type: Secondary|Time Frame: 6 months, 1 year, 2 year, 3 year, 4 year, 5 year—after the transplant

Description	No device assistance needed for ambulation evaluated at 6 months, 1 year, 2 year, 3 year, 4 year, 5 year—after the transplant
Time frame	6 months, 1 year, 2 year, 3 year, 4 year, 5 year—after the transplant
Analysis population description	The reason that the number analyzed in one or more rows differs from the overall number analyzed is because one patient with coexistent SLE died of complications from active lupus 10 months after the transplant and another patient was only 3 years out from transplant

Arm/group title	Hematopoietic stem cell transplantation
Arm/group description	Hematopoietic stem cell transplantation will be performed after conditioning regimen of cyclophosphamide, G-CSF, Mesna, rATG, rituximab, and methylprednisolone Hematopoietic Stem Cell Transplantation: Infusion of participant's own stem cells Cyclophosphamide: A medication used as chemotherapy and to suppress the immune system G-CSF: A glycoprotein that stimulates the bone marrow to produce granulocytes and stem cells and release them into the bloodstream rATG: A rabbit polyclonal antibody to lymphocytes Mesna: A medication used in those taking cyclophosphamide or ifosfamide to decrease the risk of bleeding from the bladder Rituximab: Monoclonal antibody therapy used to treat certain autoimmune diseases and types of cancer Methylprednisolone: A corticosteroid medication used to suppress the immune system and decrease inflammation
Overall number of participants analyzed	12

Pre HSCT—no assistive required	
Number analyzed	12 participants
[a]Measure type: count of participants | unit of measure: participants	650.0%
6 Mos post HSCT—no assistive device required	
Number analyzed	12 participants
[a]	975.0%
1 year post HSCT—no assistive device required	
Number analyzed	12 participants
[a]	1083.3%
2 year post HSCT—no assistive device required	
Number analyzed	12 participants
[a]	975.0%
3 year post HSCT—no assistive device required	
Number analyzed	11 participants
[a]	872.7%
4 year post HSCT—no assistive device required	
Number analyzed	12 participants
[a]	866.7%
5 year post HSCT—no assistive device required	
Number analyzed	11 participants
[a]	981.8%

[a]Measure type: count of participants|unit of measure: participants

5. Disability Score: Expanded Disability Status Scale (EDSS)
   Type: Secondary | Time Frame: pretransplant 6 month, 5 year

Description	Disability scores (disease improvement defined by at least a one point increase in the Expanded Disability Status Scale (EDSS) on consecutive evaluations at least 3 months apart The EDSS scale ranges from 0 to 10 in 0.5 unit increments that represent higher levels of disability
Time frame	Pretransplant 6 months, 5 years
Analysis population description	The reason that the number analyzed in one or more rows differs from the overall number analyzed is because one patient with coexistent SLE died of complications from active lupus 10 months after the transplant and another patient was only 3 years out from transplant

Arm/group title	Hematopoietic stem cell transplantation
Arm/group description	Hematopoietic stem cell transplantation will be performed after conditioning regimen of cyclophosphamide, G-CSF, Mesna, rATG, rituximab, and methylprednisolone Hematopoietic Stem Cell Transplantation: Infusion of participant's own stem cells Cyclophosphamide: A medication used as chemotherapy and to suppress the immune system G-CSF: A glycoprotein that stimulates the bone marrow to produce granulocytes and stem cells and release them into the bloodstream rATG: A rabbit polyclonal antibody to lymphocytes Mesna: A medication used in those taking cyclophosphamide or ifosfamide to decrease the risk of bleeding from the bladder Rituximab: Monoclonal antibody therapy used to treat certain autoimmune diseases and types of cancer Methylprednisolone: A corticosteroid medication used to suppress the immune system and decrease inflammation
Overall number of participants analyzed	13
Pretransplant disability score (EDSS)	
Number analyzed	13 participants
[a]Mean (full range)|unit of measure: score on a scale	4.4 (2.0 to 6.5)
1 year posttransplant disability score (EDSS)	
Number analyzed	12 participants
[a]	2.8 (1 to 6.5)
5 year posttransplant disability score (EDSS)	
Number analyzed	11 participants
[a]	3.3 (0 to 6.5)

[a]Mean (full range)|unit of measure: score on a scale

6. NMO-IGG Aquaporin- 4 Autoantibody Titer
   Type: Secondary | Time Frame: Pretransplant and 5 year Posttransplant

Description	NMO-IgG aquaporin- 4 autoantibody titer will be tested pretransplant and posttransplant
Time frame	Pretransplant and 5 year Posttransplant
Analysis population description	The reason that the number analyzed in one or more rows differs from the overall number analyzed is because one patient with coexistent SLE died of complications from active lupus 10 months after the transplant and another patient was only 3 years out from transplant

Arm/group title	Hematopoietic stem cell transplantation
Arm/group description	Hematopoietic stem cell transplantation will be performed after conditioning regimen of cyclophosphamide, G-CSF, Mesna, rATG, rituximab, and methylprednisolone Hematopoietic Stem Cell Transplantation: Infusion of participant's own stem cells Cyclophosphamide: A medication used as chemotherapy and to suppress the immune system G-CSF: A glycoprotein that stimulates the bone marrow to produce granulocytes and stem cells and release them into the bloodstream rATG: A rabbit polyclonal antibody to lymphocytes Mesna: A medication used in those taking cyclophosphamide or ifosfamide to decrease the risk of bleeding from the bladder Rituximab: Monoclonal antibody therapy used to treat certain autoimmune diseases and types of cancer Methylprednisolone: A corticosteroid medication used to suppress the immune system and decrease inflammation
Overall number of participants analyzed	13
Pretransplant NMO ASSAY positive	
Number analyzed	12 participants
[a]Measure type: count of participants|unit of measure: participants	1191.7%
5 year posttransplant NMO assay positive	
Number analyzed	11 participants
[a]	218.2%

[a]Measure type: count of participants | unit of measure: participants

Adverse Events

Time Frame

5 years

Adverse event reporting description

Arm/group title	Hematopoietic stem cell transplantation
Arm/group description	Hematopoietic stem cell transplantation will be performed after conditioning regimen of cyclophosphamide, G-CSF, Mesna, rATG, rituximab, and methylprednisolone Hematopoietic Stem Cell Transplantation: Infusion of participant's own stem cells Cyclophosphamide: A medication used as chemotherapy and to suppress the immune system G-CSF: A glycoprotein that stimulates the bone marrow to produce granulocytes and stem cells and release them into the bloodstream rATG: A rabbit polyclonal antibody to lymphocytes Mesna: A medication used in those taking cyclophosphamide or ifosfamide to decrease the risk of bleeding from the bladder Rituximab: Monoclonal antibody therapy used to treat certain autoimmune diseases and types of cancer Methylprednisolone: A corticosteroid medication used to suppress the immune system and decrease inflammation
All-cause mortality	
Arm/group title	Hematopoietic stem cell transplantation
	Affected/at Risk (%)
Total	1/13 (7.69%)
Serious adverse events	

Arm/group title	Hematopoietic stem cell transplantation	
	Affected/at Risk (%)	# Events
Total	1/13 (7.69%)	
Immune system disorders		
Unrelated complications from systemic lupus erythematosus[a]	1/13 (7.69%)	1
[a] Indicates events were collected by systematic assessment		
Other (not including serious) adverse events		
0% Frequency threshold for reporting other adverse events		
Arm/group title	Hematopoietic Stem Cell Transplantation	
	Affected/at Risk (%)	# Events
Total	9/13 (69.23%)	

General disorders		
Neutropenic fevers[a,b]	5/13 (38.46%)	5
Orthostatic hypotension[a,c]	2/13 (15.38%)	2
Hypertension, hyperglycemia, diarrhea[a,d]	1/13 (7.69%)	1
Infections and infestations		
Infection[a,e]	1/13 (7.69%)	1
Metabolism and nutrition disorders		
Hypophosphatemia[a,f]	9/13 (69.23%)	9
Hypocalcemia[a,g]	3/13 (23.08%)	3
Nausea and vomiting[a,h]	3/13 (23.08%)	3
Hypokalemia[a,i]	2/13 (15.38%)	2

[a]Indicates events were collected by systematic assessment
[b]Neutropenic fevers
[c]orthostatic hypotension
[d]hypertension, hyperglycemia and diarrhea
[e]C.Diff
[f]hypophosphatemia
[g]hypocalcemia
[h]nausea and vomiting
[i]hypokalemia

Completed

Study of Human Central Nervous System Stem Cells (HuCNS-SC) in Age-Related Macular Degeneration (AMD)

ClinicalTrials.gov ID NCT01632527

Sponsor: StemCells, Inc.
Information provided by StemCells, Inc. (Responsible Party)
Last Update Posted 2015-09-14

Brief Summary
The purpose of this Phase I/II study is to investigate the safety and preliminary efficacy of unilateral subretinal transplantation of HuCNS-SC cells in subjects with geographic atrophy secondary to age-related macular degeneration.

Detailed Description
This study is an open-label dose-escalation investigation of the safety and preliminary efficacy of unilateral subretinal transplantation of HuCNS-SC cells in subjects with Geographic Atrophy secondary to Age-Related Macular Degeneration (AMD). Subjects will be enrolled based on specific inclusion/exclusion criteria and evaluated at regular posttransplant intervals. The investigation will be divided into two sequential cohorts. Subjects will be enrolled into each cohort based on best-corrected visual acuity (BCVA) visual acuity as determined by the Electronic Early Treatment

Diabetic Retinopathy Study (E-ETDRS) acuity test. BCVA of less than or equal to 20/400 in the Study Eye will be enrolled in Cohort I. Subjects with BCVA of 20/320 to 20/100 in the Study Eye will be enrolled in Cohort II. Cohort I will consist of four subjects who will undergo transplant with 200,000 cells followed by four subjects who will undergo transplant with one million cells. Cohort II will consist of seven subjects who will undergo transplant with one million cells. An independent Data Monitoring Committee (DMC) will review accruing safety data for all subjects.

HuCNS-SC cells will be transplanted by a board-certified vitreoretinal surgeon. The transplantation will be conducted in the eye with the inferior best-corrected visual acuity (BCVA) (i.e., the Study Eye). Only the Study Eye will undergo transplantation. The HuCNS-SC cells will be administered into the subretinal space through a standard surgical approach.

Immunosuppressive agents will be administered orally to all subjects for a period of three months after surgery.

Subjects will be monitored frequently for a total of one year after HuCNS-SC cell transplantation. An additional four years of monitoring will be conducted in a separate follow-up study that will commence with the termination visit of the Phase I/II investigation. The follow-up study will be conducted as a separate investigation.

Official Title
Phase I/II Study of the Safety and Preliminary Efficacy of Human Central Nervous System Stem Cells (HuCNS-SC) Subretinal Transplantation in Subjects With Geographic Atrophy of Age-Related Macular Degeneration

Conditions
Age-Related Macular Degeneration
Macular Degeneration
AMD

Intervention/Treatment
Drug: HuCNS-SC cells

Study Start
2012-06

Primary Completion (Actual)
2015-06

Study Completion (Actual)
2015-06

Enrollment (Actual)
15

Study Type
Interventional

Phase
Phase 1 Phase 2

Other Study ID Numbers
StemCells CL-N01-AMD

United States
California Locations

Los Angeles, California, United States, 90,211
Retina-Vitreous Associates Medical Group

Palo Alto, California, United States, 94,303
Byers Eye Institute at Stanford, Stanford Hospital and Clinics

New York Locations

New York, New York, United States, 10,003
New York Eye and Ear Infirmary

Texas Locations

Abilene, Texas, United States, 79,606
Retina Research Institute of Texas

Dallas, Texas, United States, 75,231
Retina Foundation of the Southwest

Eligibility Criteria
Description

Inclusion Criteria
- Diagnosis of age-related macular degeneration with geographic atrophy (GA)
- Only patients with a specific degree and extent of GA will be eligible
- Subjects with best-corrected visual acuity (BCVA) of less than or equal to 20/400 in the Study Eye will be enrolled in Cohort I. Subjects with BCVA of 20/320 to 20/100 in the Study Eye will be enrolled in Cohort II
- No prior or current choroidal neovascularization in either eye
- Must have adequate care-giver support and access to medical care in the local community
- Able to provide written informed consent prior to any study-related procedures
- Agree to comply in good faith with all conditions of the study and to attend all required study visits

Exclusion Criteria
- Prior vitreal or retinal surgery
- Glaucoma
- Atrophic macular disease of any other cause
- Diabetic retinopathy or diabetic macular edema in either eye
- Previous organ, tissue, or bone marrow transplantation
- Previous participation in a gene transfer or a cell transplant trial
- Autoimmune disease
- Allergy to tacrolimus, mycophenolate mofetil (MMF), scopolamine, Cyclogyl, Moxifloxacin, or Gatifloxacin
- Current or prior malignancy (or is on chemotherapy)

Ages Eligible for Study
50 years and older (Adult, Older Adult)

Sexes Eligible for Study
All

Accepts Healthy Volunteers
No

Design Details
Primary Purpose: Treatment
Allocation: N/A
Interventional Model: Single Group Assignment
Masking: None (Open Label)

Arms and interventions

Participant group/arm	Intervention/treatment
Experimental: HuCNS-SC HuCNS-SC cells	Drug: HuCNS-SC cells • HuCNS-SC cells transplantation directly into the subretinal space of one eye (Study eye) in a single transplant procedure • Other Names: – Human Central Nervous System Stem Cells

Primary outcome measures

Outcome measure	Measure description	Time frame
Number of subjects with adverse events	Descriptive analysis of frequency and types of adverse events experienced by each subject during the study period.	One year following transplant surgery

Secondary outcome measures

Outcome measure	Measure description	Time frame
Assessment of visual function changes from baseline	Assessments will include BCVA by the ETDRS acuity test, fluorescein angiography and fundus photography, spectral domain ocular coherence tomography (OCT), microperimetry, multifocal electroretinography, contrast sensitivity, and a standardized questionnaire of visual function	At frequent intervals for one year following transplantation

Sponsor
StemCells, Inc.

Collaborators
No information provided

Investigators
- Study Director: Stephen Huhn, M.D., StemCells, Inc.

General Publications
No publications available

Brazil

Completed

Autologous Bone Marrow-Derived Stem Cells Transplantation For Retinitis Pigmentosa (RETICELL)

ClinicalTrials.gov ID NCT01560715

Sponsor: University of Sao Paulo
Information provided by Rubens Camargo Siqueira, University of Sao Paulo (Responsible Party)
Last Update Posted 2017-07-21

Brief Summary
The purpose of this study is to evaluate the short-term safety and efficacy of a single intravitreal injection of autologous bone marrow stem cells in patients with retinitis pigmentosa.

Detailed Description
A prospective phase II, non-randomized open-label study of retinitis pigmentosa patients with best-corrected ETDRS visual acuity (BCVA) worse than 20/200. Standardized ophthalmic evaluation will be perform at baseline and at weeks 1, 4, 12, and 24 (±1) following intravitreal injection of 10 × 106 bone marrow stem cells/0.1 ml. Three measures will be used to evaluate the short-term safety of intravitreal of ABMDSC: (1) severe visual loss, defined as a drop in 15 letters on ETDRS visual acuity scale; (2) decrease in ERG response; (3) decrease in 5 square degrees on visual field; secondary safety outcomes: (1) increase in intra-ocular inflammation defined herein as anterior chamber cells and flare higher than 3+ for more than 1 month after injection according to a classification described elsewhere; (2) decrease in CMT more than 50 um; (3) genesis of abnormal tissues (teratomas) or tumors; (4) qualitative changes in retinal or choroidal perfusion, like macular non-perfusion. Secondary outcome measures will be used to evaluate the short-term efficiency of intravitreal of ABMDSC: (1) improvement in ERG response; (2) increase in visual field; (3) increase in CSMT >50 um and not related to macular edema; (4) increase >5 letters on BCVA.

Official Title
Phase 2 Study Of Autologous Bone Marrow-Derived Stem Cells Transplantation For Retinitis Pigmentosa

Conditions
Retinitis Pigmentosa

Study Type
Interventional

Enrollment (Actual)
50

Intervention/Treatment
Biological: intravitreal injection of autologous bone marrow stem cells

Phase
Phase 2

Other Study ID Numbers
Rubens Siqueira

Study Start
2011-06

Primary Completion (Actual)
2013-01

Study Completion (Actual)
2013-06

Brazil
Locations

Sao Jose do Rio Preto, SP, Brazil, 15010-100
Centro de Pesquisa Rubens Siqueira

Eligibility Criteria
Ages Eligible For Study
18 to 75 years (Adult, Older Adult)

Accepts Healthy Volunteer
No

Sexes Eligible For Study
All

Description

Inclusion Criteria
- diagnosis of retinitis pigmentosa logarithm of minimum angle of resolution (logMAR) BCVA of 1.0 (Snellen equivalent, 20/200) or worse
- visual Field less than 20°

Exclusion Criteria
- previous ocular surgery other than cataract
- presence of cataract or other media opacity that would influence ocular fundus documentation and adequate ERG and visual field evaluation
- other ophthalmic disease like glaucoma and uveitis
- previous history of blood disorders like leukemia
- known allergy to fluorescein or indocyanine green
- known coagulation abnormalities or current use of anticoagulative medication other than aspirin.

Design Details

Primary Purpose: Treatment
Allocation: N/A
Interventional Model: Single Group Assignment
Masking: None (Open Label)

Arms and interventions

Participant group/arm	Intervention/treatment
Experimental: Experimental: Test group Retinitis pigmentosa patients with best-corrected visual acuity (BCVA) worse than 20/200 or visual field less than 20°	Biological: intravitreal injection of autologous bone marrow stem cells • One intravitreal injection of a 0.1-ml cell suspension containing around 10 × 106 bone marrow mononuclear stem cells (BMMSC). All treatments were performed by a single retinal specialist using topical proparacaine drops under sterile conditions (eyelid speculum and povidone-iodine). Autologous BMMSC were injected into the vitreous cavity using a 27 gauge needle inserted through the inferotemporal pars plana 3.0–3.5 mm posterior to the limbus. After the injection, central retinal artery perfusion was confirmed with indirect ophthalmoscopy. Patients were instructed to instill one drop of 0.3% ciprofloxacin into the injected eye four times daily for 1 week after the procedure • Other Names: – autologous bone marrow stem cells

Primary outcome measures

Outcome measure	Measure description	Time frame
ETDRS visual acuity change		Day 1 to Day 365

Sponsor

University of Sao Paulo

Collaborators

No information provided

Investigators

- Principal Investigator: Rubens C Siqueira, MD, PhD, Sao Paulo University

General Publications

- Siqueira RC, Messias A, Voltarelli JC, Scott IU, Jorge R. Intravitreal injection of autologous bone marrow-derived mononuclear cells for hereditary retinal dystrophy: a phase I trial. Retina. 2011 Jun;31(6):1207-14. https://doi.org/10.1097/IAE.0b013e3181f9c242.
- Siqueira RC, Messias A, Messias K, Arcieri RS, Ruiz MA, Souza NF, Martins LC, Jorge R. Quality of life in patients with retinitis pigmentosa submitted to intravitreal use of bone marrow-derived stem cells (Reticell -clinical trial). Stem Cell Res Ther. 2015 Mar 14;6(1):29. https://doi.org/10.1186/s13287-015-0020-6.
- Siqueira RC. Stem cell therapy in retinal diseases? Rev. Bras Hematol Hemoter. 2012;34(3):222-6. https://doi.org/10.5581/1516-8484.20120054. No abstract available.
- Siqueira RC, Messias A, Voltarelli JC, Messias K, Arcieri RS, Jorge R. Resolution of macular oedema associated with retinitis pigmentosa after intravitreal use of autologous BM-derived hematopoietic stem cell transplantation. Bone Marrow Transplant. 2013 Apr;48(4):612-3. https://doi.org/10.1038/bmt.2012.185. Epub 2012 Sep 24. No abstract available.

Completed

Autologous Bone Marrow-Derived Stem Cells Transplantation For Retinitis Pigmentosa

ClinicalTrials.gov ID NCT01068561

Sponsor: University of Sao Paulo
Information provided by Rubens Camargo Siqueira, University of Sao Paulo (Responsible Party)
Last Update Posted 2011-09-20

Brief Summary
The purpose of this study is to evaluate the short-term safety of a single intravitreal injection of autologous bone marrow stem cells in patients with retinitis pigmentosa.

Detailed Description
A prospective phase I, non-randomized open-label study of retinitis pigmentosa patients with best-corrected ETDRS visual acuity (BCVA) worse than 20/200. Standardized ophthalmic evaluation will be performed at baseline and at weeks 1, 4, 12, and 24 (±1) following intravitreal injection of 10 × 106 bone marrow stem cells/0.1 ml. Three measures will be used to evaluate the short-term safety of intravitreal of ABMDSC: (1) severe visual loss, defined as a drop in 15 letters on ETDRS visual acuity scale; (2) decrease in ERG response; (3) decrease in 5 square degrees on visual field; secondary safety outcomes: (1) increase in intra-ocular inflammation defined herein as anterior chamber cells and flare higher than 3+ for more than 1 month after injection according to a classification described elsewhere; (2) decrease in CMT more than 50 um; (3) genesis of abnormal tissues (teratomas) or tumors; (4) qualitative changes in retinal or choroidal perfusion, like macular nonperfusion. Secondary outcome measures will be used to evaluate the short-term efficiency of intravitreal of ABMDSC: (1) improvement in ERG response; (2) increase in visual field: (3) increase in CSMT >50 um and not related to macular edema; (4) increase >5 letters on BCVA.

Official Title
Phase 1 Study Of Autologous Bone Marrow-Derived Stem Cells Transplantation For Retinitis Pigmentosa

Conditions
Retinitis Pigmentosa

Study Type
Interventional

Enrollment (Actual)
5

Intervention/Treatment
Biological: intravitreal injection of autologous bone marrow stem cells

Phase
Phase 1

Other Study ID Numbers
CPRS

Study Start
2009-05

Primary Completion (Actual)
2010-01

Study Completion (Actual)
2010-02

Brazil

São Jose do Rio Preto, SP, Brazil, 15010-100

Eligibility Criteria
Ages Eligible for Study
18 years and older (Adult, Older Adult)

Accepts Healthy Volunteer
No

Sexes Eligible for Study
All

Description

Inclusion Criteria
- diagnosis of retinitis pigmentosa
- logarithm of minimum angle of resolution (logMAR) BCVA of 1.0 (Snellen equivalent, 20/200) or worse

Exclusion Criteria
- previous ocular surgery other than cataract
- presence of cataract or other media opacity that would influence ocular fundus documentation and adequate ERG and visual field evaluation
- other ophthalmic disease like glaucoma and uveitis
- previous history of blood disorders like leukemia
- known allergy to fluorescein or indocyanine green
- known coagulation abnormalities or current use of anticoagulative medication other than aspirin.

Design Details
Primary Purpose: Treatment
Allocation: Non-Randomized
Interventional Model: Single Group Assignment
Masking: None (Open Label)

Arms and interventions

Participant group/arm	Intervention/treatment
Experimental: Test group open-label study of retinitis pigmentosa patients with best-corrected visual acuity (BCVA) worse than 20/200	Biological: intravitreal injection of autologous bone marrow stem cells • One intravitreal injection of a 0.1-ml cell suspension containing around 10 × 106 bone marrow mononuclear stem cells (BMMSC). All treatments were performed by a single retinal specialist using topical proparacaine drops under sterile conditions (eyelid speculum and povidone-iodine). Autologous BMMSC were injected into the vitreous cavity using a 27 gauge needle inserted through the inferotemporal pars plana 3.0–3.5 mm posterior to the limbus. After the injection, central retinal artery perfusion was confirmed with indirect ophthalmoscopy. Patients were instructed to instill one drop of 0.3% ciprofloxacin into the injected eye four times daily for 1 week after the procedure • Other Names: – Autologous bone marrow mononuclear stem cells

Primary outcome measures

Outcome measure	Measure description	Time frame
Severe visual loss, defined as a drop in 15 letters on ETDRS visual acuity scale		3–6 months

Sponsor
University of Sao Paulo

Collaborators
No information provided

Investigators
- Principal Investigator: Rubens C Siqueira, MD, PhD, Research Center Rubens Siqueira, Catanduva Medicine School and São Paulo University

General Publications
No publications available

Completed

Stem Cell Therapy for Outer Retinal Degenerations

ClinicalTrials.gov ID NCT02903576

Sponsor: Federal University of São Paulo
Information provided by Rubens Belfort Jr., Federal University of São Paulo (Responsible Party)
Last Update Posted 2020-07-22

Brief Summary
This study is a Phase I/II, open-label, non-randomized, prospective study to determine the safety of human embryonic stem cell-derived Retinal pigmented epithelium (hESC RPE) subretinal injections versus hESC RPE seeded on a polymeric substrate implanted in the subretinal space.

Detailed Description
To determine whether the surgical implantation of a human embryonic stem cell-derived retinal pigmented epithelium (hESC-RPE) monolayer seeded onto a polymeric versus hESC-RPE injections into the subretinal space is a safe procedure.

Six patients will receive hESC-RPE cell injections (100,000 cells) in the subretinal space (2 Dry Age-related macular degeneration (AMD), 2 Wet AMD with disciform scar, and 2 with Stargardt disease).

Also five patients with Dry AMD, five patients with Wet AMD with disciform scar, and five patients with Stargardt disease with receive a subretinal implantation of the hESC-RPE seeded in a monolayer in a polymeric substrate.

Patients will be enrolled sequentially, and after the procedure, the patients will be followed for 1 year.

Official Title
Stem Cell-Derived Retinal Pigmented Epithelium Implantation in Patients With Outer Retinal Degenerations: Phase I/II Clinical Trial

Conditions
Age-Related Macular Degeneration
Exudative Age-related Macular Degeneration

Study Type
Interventional

Enrollment (Actual)
15

Intervention/Treatment
Procedure: injection of hESC-RPE in suspension
Procedure: injection hESC-RPE seeded in a substrate

Phase
Phase 1 Phase 2

Other Study ID Numbers
12018712.5.0000.5505

Study Start
2015-08

Primary Completion (Actual)
2019-01

Study Completion (Actual)
2019-06

Brazil

Eligibility Criteria
Ages Eligible for Study
18 to 90 years (Adult, Older Adult)

Accepts Healthy Volunteer
No

Sexes Eligible For Study
All

Description

Inclusion Criteria
- Patients with AMD (Dry, Wet after failure of treatment, disciform scars)
- Patients with Stargardt Disease BCVA on the selected eye: worse than 20/200

Exclusion Criteria
- Other ophthalmological diseases (Glaucoma, Diabetic Retinopathy, Previous retinal surgery, uveitis)
- Systemic diseases with contraindication for surgical procedures with local anesthesia

Design Details
Primary Purpose: Treatment
Allocation: Non-Randomized
Interventional Model: Parallel Assignment
Masking: None (Open Label)

Arms and interventions

Participant group/arm	Intervention/treatment
Active Comparator: injection of hESC-RPE in suspension Six patients will receive cell suspension injections on the subretinal space prior to the surgeries, to access safety	Procedure: injection of hESC-RPE in suspension • The first six patients will receive a subretinal injection of hESC-RPE in solution after a complete pars plana vitrectomy to access safety of the cell implant alone. • Other Names: – Human Embryonic Stem Cell—Retinal Pigmented Epithelium
Active Comparator: injection hESC-RPE seeded in a substrate 15 patients will receive a subretinal implantation of a polymeric scaffold seeded hESC-RPE in monolayer	Procedure: injection hESC-RPE seeded in a substrate • Fifteen patients will receive a subretinal implantation of embryonic stem cell-derived retinal pigmented epithelium seeded in a polymeric substrate to access safety of substrate seeded with RPE • Other Names: – Human Embryonic Stem Cell seeded in a polymeric substrate

Primary outcome measures

Outcome measure	Measure description	Time frame
Incidence of Treatment-Emergent Adverse Events of subretinal implantation of stem cell-derived retinal pigmented epithelium in the subretinal space	Incidence of surgical-related side effects: Retinal detachment, Ocular inflammation, Increase in intraocular pressure, Infection (endophthalmitis), Loss of vision due to surgical-related complications	1 year

Secondary outcome measures

Outcome measure	Measure description	Time frame
Incidence of side effects related to the treatment itself (injection and implantation of subretinal stem cell-related RPE)	Inflammation/rejection Cell migration/ differentiation Tumor formation Proliferative vitreoretinopathy/Retinal detachment Implant migration	1 year

Sponsor

Federal University of São Paulo

Collaborators

No information provided

Investigators

- Study Chair: Rubens Belfort, MD, Federal University of São Paulo UNIFESP

General Publications

No publications available

Turkey

Completed

Spheroidal Mesenchymal Stem Cells in Retinitis Pigmentosa

ClinicalTrials.gov ID NCT05712148

Sponsor: Acibadem University
Information provided by Berna Özkan, Acibadem University (Responsible Party)
Last Update Posted 2023-02-03

Brief Summary

Purpose: This prospective clinical case series aimed to evaluate the effect of suprachoroidal implantation of mesenchymal stem cells (MSCs) in the form of spheroids as a stem cell therapy for retinitis pigmentosa (RP) patients with relatively good visual acuity.

Methods: Fifteen eyes of 15 patients with RP who received suprachoroidal implantation of MSCs in the form of spheroids were included. Best-corrected visual acuity (BCVA), 10-2 and 30-2 visual field examination, and multifocal electroretinography (mfERG) recordings were recorded at baseline, postoperative first, third-, and sixth-months during follow-up.

Detailed Description

A prospective clinical trial was conducted in patients with RP at the department of ophthalmology, Acibadem University, Medical school. The study was approved by the Review Board of Cell, Organ and Tissue Transplantation, Department of Turkish Ministry of Health. The study was performed in accordance with the Declaration of Helsinki.

Patient Evaluation: Fifteen patients with clinical and genetic diagnoses of RP were included in the study. Patients underwent ophthalmic examination including best-corrected visual acuity (BCVA), intraocular pressure, anterior segment examination, and dilated fundus examination (with topical tropicamide %1 and phenylephrine 2.5%). Each eye underwent spectral domain OCT scanning with Cirrus 5000 HD-OCT Angioplex (Carl Zeiss Meditec, Dublin, CA, USA), fundus autofluorescence, and fundus fluorescein angiography (Heidelberg Engineering, Germany). MD (mean deviation) and PSD (pattern standard deviation) parameters of 10-2 and 30-2 visual field (VF) testing strategies with a Humphrey Field Analyzer model 750I (Carl Zeiss Meditec, Dublin, CA, USA) were obtained. The electrophysiological function was assessed with mfERG evaluation (Monpack 3, Metrovision, France) according to the International Society for Clinical Electrophysiology of Vision (ISCEV) guidelines. BCVA was converted to the logarithm of the minimal angle of resolution (logMAR) equivalent.

The patients were excluded from the study, if they had (1) coexisting ocular pathology that may affect visual acuity, visual field, and retinal morphology such as glaucoma, uveitis, previous vitreoretinal surgery; (2) coexisting cataract that may affect mfERG, visual field, and/or ocular imaging; (3) refractive error that may affect measurements higher than +6.00D, lower than −6.00D; (4) coexisting systemic diseases that may affect visual function such as diabetes, vasculitis, rheumatological diseases, and chronic immunosuppressive use; (5) periocular injection of platelet-rich blood and transcorneal electrical stimulation in previous 6 months and (6) previous ocular surgery.

Electrophysiologic Testing: After 30 min of dark adaptation and pupil dilatation with the application of one drop of tropicamide 1% (Tropamid, Bilim ˙Ilaç, Turkey), phenylephrine 2.5% (Mydfrin, Alcon), and proparacaine hydrochloride 0.5% (Alcaine, Alcon), ERG jet electrodes were placed. Multifocal electroretinographies were recorded after pupil dilatation. The stimulated retinal area was subtended in an area of 60° × 55°; 61 hexagon stimulants were used with alternating black (5 cd/m^2) and white (100 cd/m^2) stimulants. The concentric rings were analyzed according to International Society for Clinical Electrophysiology of Vision standards.13 The amplitude and latencies of P1, N1, and N2 components were recorded for every ring. The mean signal amplitudes (MSAs) of multifocal electroretinography (mfERG) in the macula (central 0°–2°) and the peripheral (2°–5°,5°–10°, 10°–15°, and >15°) signal amplitude changes were evaluated separately.

Spheroidal Stem Cell Preparation: First passage umbilical cord-derived mesenchymal stem cell was obtained from Labcell Cellular Laboratory Center which provides GMP (good manufacturing practices) conditions.

Spheroid Production: The spheroid production continued in GMP condition. First passage umbilical cord-derived mesenchymal stem cells were used in the production of spheroids. 100,000 mesenchymal stem cells were suspended with 100 μl. Serum-free medium (MSC NutriStem® XF Medium, Sartorius) containing %1 ciprofloxacin (Polipharma). Each well of the low-attachment 96 well plate was seeded with 100,000 cells in 100 μl. medium. The cells were incubated at 37 °C for 48 h. At the end of 48 h, all spheroids were collected with a micropipette and transferred into Ringer's lactate solution (Osel/Biofleks) containing 1% HSA (csl behring). Spheroids were washed 3 times with Ringer's lactate solution containing 1% HAS (csl behring). 50 spheroids were produced for one patient (50 spheroids containing 5 × 106 cells were produced). 5 of 50 spheroids were reserved for quality control analysis. The remaining 45 spheroid membranes were embedded in the matrix.

Matrix Production and Cell Embedding Culture: Matrix mixture containing 225 μl cryoprecipitate +22.5 μl calcium (Adeka) + 2.5 μl transamine (Haver) was added to each well of the 96 well plates. When the matrix became semisolid, 45 spheroids were embedded in the middle of the matrix and incubated at 37 °C for 45 min. The matrix, which became completely solid after 45 min, was removed with the help of a scalpel, transferred into Ringer's lactate solution (Osel/Biofleks) containing 1% HAS (csl behring), and transferred to the operating room in this solution.

Quality Control Analysis Microbiological blood culture, fungal, endotoxin analysis and purity (ciprofloxacin <0.1 μg/ml), efficiency (adipocyte and cartilage differentiation analysis) Cell Count/Viability and flow cytometric analysis (CD34 < %2,

CD45 < %4, CD90 > %80, HLA-DR < %4, CD105 > %60, CD73 > %70) were studied from the reserved spheroids for quality control analyses.

Surgical Technique: Patients underwent surgery under general anesthesia. The inferotemporal quadrant was chosen as the surgical area. 6.0 silk was used as the anchoring suture near the limbus between 4 and 5 o'clock. Conjunctiva and tenon were opened as 6 mm long cut at 6 mm from limbus parallel to the limbus and the edges of the cut were advanced 3 mm posteriorly. Two 8.0 vicryl sutures were used as traction sutures at the anterior corners of the conjunctiva. The tenon was dissected over the sclera posteriorly. Then, we performed a 7 × 7 scleral flap. Anterior margin of the scleral flap was created at 8 mm from the limbus, parallel to the limbus with 30° ophthalmic knife. Two other half-thickness side cuts were made to create a U-shaped flap that has its base parallel to the lateral rectus muscle. Then, starting from the infero-anterior edge, a deep scleral flap was dissected with a crescent blade. During the dissection, black choroidal reflex should be observed all around the surface of the flap bed. The fibrin plug carrying spheroidal stem cells was placed over the choroid that was seen under the thin sclera. It was covered by the scleral flap and the flap was sutured to its original position from its edges with a 7.0 vicryl suture. Tenon and conjunctiva were closed separately with an 8.0 vicryl suture.

Official Title
Suprachoroidal Spheroidal Mesenchymal Stem Cell Implantation in Retinitis Pigmentosa: Clinical Results of 6 Months Follow-up

Conditions
Retinitis Pigmentosa
Stem Cell Transplant

Study Type
Interventional

Enrollment (Actual)
15

Intervention/Treatment
Biological: Stem cell implantation

Phase
Phase 1 Phase 2

Other Study ID Numbers
56733164/203 E.3858

Study Start (Actual)
2019-10-11

Primary Completion (Actual)
2022-01-30

Study Completion (Actual)
2022-06-14

Istanbul, Turkey
Acıbadem University

Eligibility Criteria
Ages Eligible For Study
(Child, Adult, Older Adult)

Accepts Healthy Volunteer
No

Sexes Eligible for Study
All

Description

Inclusion Criteria
Clinical and genetic diagnosis of Retinitis Pigmentosa

Exclusion Criteria
- Coexisting ocular pathology that may affect visual acuity, visual field, and retinal morphology such as glaucoma, uveitis, previous vitreoretinal surgery
- Coexisting cataract that may affect mfERG, visual field. and/or ocular imaging
- Refractive error that may affect measurements higher than +6.00D, lower than −6.00D
- Coexisting systemic diseases that may affect visual function such as diabetes, vasculitis, rheumatological diseases, and chronic immunosuppressive use
- Periocular injection of platelet-rich blood and transcorneal electrical stimulation in previous 6 months and (6) previous ocular surgery

Design Details
Primary Purpose: Supportive Care
Allocation: N/A
Interventional Model: Single Group Assignment
Masking: None (Open Label)

Arms and interventions

Participant group/arm	Intervention/treatment
Experimental: Retinitis Pigmentosa patients Suprachoroidal mesenchymal stem cell-implanted Retinitis pigmentosa patients	Biological: Stem cell implantation • Suprachoroidal mesenchymal spheroidal stem cell implantation

Primary outcome measures

Outcome measure	Measure description	Time frame
Change in visual acuity after stem cell implantation in Retinitis Pigmentosa	Preimplantation and postimplantation BCVA was converted to the logarithm of the minimal angle of resolution (logMAR) equivalent	3 years
Change in microperimetric evaluation after stem cell implantation in Retinitis Pigmentosa	MD (mean deviation) and PSD (pattern standard deviation) parameters of 10-2 and 30-2 visual field (VF) testing strategies with a Humphrey Field Analyzer model 750I (Carl Zeiss Meditec, Dublin, CA, USA) were obtained	3 years

Change in electrophysiological testing after stem cell implantation in Retinitis Pigmentosa	The electrophysiological function was assessed with mfERG evaluation (Monpack 3, Metrovision, France) according to the International Society for Clinical Electrophysiology of Vision (ISCEV) guidelines. After 30 min of dark adaptation and pupil dilatation with the application of one drop of tropicamide 1% (Tropamid, Bilim ˙Ilaç, Turkey), phenylephrine 2.5% (Mydfrin, Alcon), and proparacaine hydrochloride 0.5% (Alcaine, Alcon), ERG jet electrodes were placed. Multifocal electroretinographies were recorded after pupil dilatation. The concentric rings were analyzed according to International Society for Clinical Electrophysiology of Vision standards. The amplitude and latencies of P1, N1, and N2 components were recorded for every ring. The mean signal amplitudes (MSAs) of multifocal electroretinography (mfERG) in the macula (central 0°–2°) and the peripheral (2°–5°, 5°–10°, 10°–15°, and >15°) signal amplitude changes were evaluated separately	3 years

Sponsor
Acibadem University

Collaborators
No information provided

Investigators
- Principal Investigator: Berna Özkan, Professor, Acibadem University
- Study Chair: Ercüment Ovalı, Professor, Acıbadem LabCell Cellular Therapy Center

General Publications
No publications available

Results Overview
No Study Results Posted on ClinicalTrials.gov for this Study

Completed

Therapy of Toxic Optic Neuropathy Via Combination of Stem Cells With Electromagnetic Stimulation (Magnovision)

ClinicalTrials.gov ID NCT04877067

Sponsor: Ankara Universitesi Teknokent
Information provided by Umut Arslan, Ankara Universitesi Teknokent (Responsible Party)
Last Update Posted 2021-05-10

Brief Summary

The axons of the retinal ganglion cells combine to form the optic nerve. The optic nerve transmits electrical signals to the visual cortex by various synapses. Optic nerve axons are more sensitive to toxins than retina because they are outside the blood retinal barrier. Methanol, various solvents and heavy metals, carbon dioxide, antiarrhythmic, antiepileptic, antibiotics, and some vasoactive drugs can cause toxic optic neuropathy. There is a different pathophysiology for each toxin. Methanol is easily accessible alcohol in all types of disinfectants. Methanol is converted into formaldehyde and formic acid while metabolized in the liver. Formaldehyde disrupts ATP synthesis by blocking mitochondrial function and oxidative phosphorylation. Formic acid causes demyelination as a result of metabolic acidosis. Neuroinflammation occurs when denatured proteins block axoplasmic flow. All these processes can lead to apoptosis and permanent vision loss. Sildenafil is a vasoactive drug used in erectile dysfunction. Sildenafil decreases optic nerve head blood flow. Neuroinflammation develops secondary to the cessation of axoplasmic flow after hypoxia. If hypoxia and neuroinflammation persists, apoptosis and permanent vision loss develop. Amiodarone is an ion channel blocker used in the treatment of cardiac arrhythmias. Long-term use may cause disruption of ion channel balance in the optic nerve. This condition leads to asymmetric neuroinflammation and apoptosis.

Wharton's jelly-derived mesenchymal stem cells (WJ-MSC) can increase mitochondrial ATP synthesis with paracrine effects and suppress neuroinflammation with immunomodulatory effects. Repetitive electromagnetic stimulation (rEMS) can rearrange ion channel balances and axoplasmic flow. The aim of this prospective phase-3 clinical study is to investigate the effect of WJ-MSC and rEMS combination in the therapy of toxic optic neuropathies. This combination is the first study in the literature for the therapy of toxic optic neuropathies.

Detailed Description

This prospective, open-label clinical phase-3 study was conducted at Ankara University Faculty of Medicine, Department of Ophthalmology between April 2019 and April 2021. Thirty-six eyes of 18 patients with toxic optic neuropathy were included in the study. The primary toxic optic neuropathy (TON) diagnosis was made in emergency or intensive care clinics. Patients who admitted within 3 months after the first intervention were included in the study. All patients enrolled underwent a complete routine ophthalmic examination, including the best-corrected visual acuity (BCVA) measurement with the early treatment of diabetic retinopathy study (ETDRS) chart (Topcon CC 100 XP, Japan). The patients were further evaluated with optical coherence tomography angiography (OCTA) from RTVue XR (Avanti, Optovue, Fremont, CA, USA) to analyze and measure the changes in the ganglion cell complex (GCC) layer that provided a typical co-registered en-face and cross-sectional multimodal imaging platform. Functional evaluation of optic nerve was followed by Compass 24/2 visual field (VF) test (Compass, CenterVue, Padova, Italy) and 120′ pattern visual evoked potential (pVEP) test (Mon 2018F, Metrovision, Perenchies, France).

Subjects: Thirty-six eyes of 18 TON patients were included in the study.

Inclusion criteria consisted of several parameters:

- Cases poisoned with a toxic agent within 3 months
- Patients with BCVA better than 35 letters
- Any degree of visual field loss
- Patients over 18 years old

Exclusion criteria consisted of several parameters:

- Cases poisoned with a toxic agent more than 3 months
- Patients with BCVA less than 35 letters: to perform the visual field test correctly
- Neurological sequelae that cannot be cooperated
- Smoking

The same treatment protocol was applied to all patients with methanol intoxication before admission to our clinic. Hemodialysis, bicarbonate, and ethanol were administered to all patients. The cases were divided into three groups according to similar demographic characteristics.

Group 1: Consists of 12 eyes of 12 patients treated with WJ-MSC and rEMS combination in one eye. WJ-MSC was applied only one time before rEMS applications. rEMS application was started 10 days after WJ-MSC application. The rEMS was applied with a custom-designed helmet for 30 min after the WJ-MSC application. rEMS applications were repeated 10 times with a 1-week interval. The course was evaluated by comparing the BCVA, FPDI, GCC thickness, pVEP-p100 latency and amplitude parameters before and third month of applications.

Group 2: Consists of 12 eyes of 12 patients treated with only rEMS in one eye. The rEMS was applied with a custom-designed helmet for 30 min. rEMS applications were repeated 10 times with a 1-week interval. The course was evaluated by comparing the BCVA, FPDI, GCC thickness, pVEP-p100 latency and amplitude parameters before and third month of applications.

Group 3: Consists of 12 eyes of 6 patients treated with only WJ-MSC in two eyes. WJ-MSC was applied only one time for both eyes. The course was evaluated by comparing the BCVA, FPDI, GCC thickness, pVEP-p100 latency and amplitude parameters before and third month of applications.

Umbilical cord Wharton's jelly-derived mesenchymal stem cells preparation: The mesenchymal stem cells used in this study were isolated from Wharton's jelly of the umbilical cord that was collected allogenicity from a single donor with the mother's consent. The umbilical cord sample was treated following several steps. Briefly, cord tissue was washed twice with phosphate-buffered saline (Lonza, Switzerland) and the Wharton's jelly part was minced using forceps and a scalpel. Minced pieces were cultivated in a cell culture dish (Greiner Bio-One, Germany) with Dulbecco's modified Eagle's medium F12 (DMEM)-low glucose with no L-glutamine (Bilogical Industries, Israel) and 10% human AB serum (Capricorn, Germany), 1% 10.000 U/ml penicillin, and 10.000 µg/ml streptomycin (Gibco, USA). All cell preparations and cultivation procedures were conducted in a current

Good Manufacturing Practice (cGMP)-accredited laboratory (Onkim Stem Cell Technologies, Turkey). The culture-expanded cells were cryopreserved at P3 using standard cryopreservation protocols until used in the following experiment. CryoSure-DEX40 (WAK-Chemie Medical, Germany) containing 55% Dimethyl Sulfoxide and 5% Dextran 40 was used as cryopreservant. The cells were characterized at the time of cryopreservation using flow cytometric analysis to determine the expression of the positive cluster of differentiation (CD) surface markers, CD90, CD105, CD73, CD44, CD29, and negative for CD34, CD45, and CD11b. Using real-time polymerase chain reaction (qPCR), the expressions of several genes, such as tumor necrosis alpha (TNF alpha) and vimentin (VIM), were analyzed. Additionally, quality control analyses, such as mycoplasma and endotoxin analyses (using the PCR and LAL test combined with sterility analysis, respectively), were also completed. Cells were solubilized from cryopreservation before being prepared for injection. Average cell viability for each treatment was over 90.0%, and each patient received 2–6 × 106 cells in a 1.5 ml saline solution.

Injection of umbilical cord WJ-MSCs: The WJ-MSCs suspension from the culture was delivered to the operating room by cold chain and used within 24 h. A total of 1.5 ml of the WJ-MSC suspension was immediately injected into the subtenon space of each eye. The procedure was conducted under topical anesthesia with proparacaine hydrochloride drops (Alcaine, Alcon, USA) and sterile conditions. 5/0 atraumatic traction suture was applied to the limbus for easy access and manipulation to application area. Postoperatively, loteprednol and tobramycin combination eye drops were given four times per day for one week, and oral amoxicillin clavulonate (1 g) was given twice a day for 5 days.

Retinal repetitive electromagnetic stimulation (rEMS): The rEMS helmet (MagnovisionTM, Bioretina Biotechnology, Ankara, Turkey) stimulated the retina and visual pathways with an electromagnetic field strength of 2000 miligauss, frequency of 42 hertz, and duration 30 min. These values were previously determined to be effective for other clinical and preclinical studies.

Timeframe

The patients were evaluated at several study timepoints:

- T0 (baseline): immediately before the applications
- T1: 1st month of applications
- T2: third month of applications

Primary Outcome Measure

- ETDRS visual acuity: The visual acuity scores obtained from the T0 and T2 examinations were analyzed and compared using statistical tests to determine effectiveness.

Secondary Outcome Measures

- Visual field sensitivity: Fundus perimetry deviation index (FPDI, %) was examined in the 24/2 visual field of the computerized perimetry records. The FPDI offers data explaining how many of the 100 flashing points and what percentage of the visual field could be correctly seen by the patient. For VF analysis, in order to avoid mistakes during the test, practice rounds were carried out three times before applications.

- Ganglion cell complex thickness (GCC thickness, μm): GCC is the thickness from the internal limiting membrane to the inner plexiform layer in the 3 × 3 mm of foveal area. The measurement done automatically by the OCTA device. GCC is the total thickness of the ganglion cells and retinal nerve fibers.
- Pattern visual evoked potential (pVEP): pVEP is an objective test that measures the electrical activity of the optical pathway in response to a light stimulus. The 120′ patterns reveals responses from the all retinal quadrants. The measurements were taken according to the ISCEV standards for both eyes. We used the 120’pattern VEP protocol, which combines p100 implicit time and amplitude, to create a numerical result.

Official Title
Bioretina Inc. Ankara University Technopolis

Conditions
Methanol Poisoning
Toxic Optic Neuropathy
Stem Cell Tyrosine Kinase 1 Y842X
Magnetic Field Exposure

Study Type
Interventional

Enrollment (Actual)
18

Intervention/Treatment
Biological: Wharton’s jelly-derived mesenchymal stem cell
Device: Repetitive electromagnetic stimulation

Phase
Phase 3

Study Start (Actual)
2019-04-01

Primary Completion (Actual)
2021-04-30

Study Completion (Actual)
2021-04-30

Turkey

Ankara, Turkey, 06000
Umut Arslan

Türkiye Locations

Ankara, Türkiye, Turkey, 06312
Ankara University Biotechnology Institute

Eligibility Criteria
Ages Eligible For Study
18 years and older (Adult, Older Adult)

Accepts Healthy Volunteer
No

Sexes Eligible For Study
All

Inclusion Criteria
- Cases poisoned with a toxic agent within 3 months
- Patients with BCVA better than 35 letters
- Any degree of visual field loss
- Patients over 18 years old

Exclusion Criteria
- Cases poisoned with a toxic agent more than 3 months
- Patients with BCVA less than 35 letters: to perform the visual field test correctly
- Neurological sequelae that cannot be cooperated
- Smoking

Primary Purpose: Treatment
Allocation: Non-Randomized
Interventional Model: Parallel Assignment
Interventional Model Description: The cases were divided into 3 groups according to their similar demographic characteristics.
Prospective open label
Masking: None (Open Label)

Arms and interventions

Participant group/arm	Intervention/treatment
Active Comparator: WJ-MSC combined with rEMS WJ-MSC was applied first to the patients after necessary preparations. rEMS application was started 10 days after WJ-MSC application	Biological: Wharton's jelly-derived mesenchymal stem cell • WJ-MSC • Other Names: – CordoSight Device: Repetitive electromagnetic stimulation • rEMS • Other Names: – Magnovision

Active Comparator: Only rEMS rEMS applications were repeated 10 times with a 1-week interval	Device: Repetitive electromagnetic stimulation • rEMS • Other Names: – Magnovision
Active Comparator: Only WJ-MSC WJ-MSC was applied only one time for both eyes	Biological: Wharton's jelly-derived mesenchymal stem cell • WJ-MSC • Other Names: – CordoSight

Primary outcome measures

Outcome measure	Measure description	Time frame
ETDRS visual acuity:	The visual acuity scores obtained from the T0 and T2 examinations were analyzed and compared using statistical tests to determine effectiveness	Change from baseline BCVA at 3 months

Secondary outcome measures

Outcome measure	Measure description	Time frame
Visual field sensitivity: Fundus perimetry deviation index (FPDI, %)	FPDI was examined in the 24/2 visual field of the computerized perimetry records. The FPDI offers data explaining how many of the 100 flashing points and what percentage of the visual field could be correctly seen by the patient. For VF analysis, in order to avoid mistakes during the test, practice rounds were carried out three times before applications	Change from baseline FPDI at 3 months
Ganglion cell complex thickness (GCC thickness, μm):	GCC is the thickness from the internal limiting membrane to the inner plexiform layer in the 3 × 3 mm of foveal area. The measurement done automatically by the OCTA device. GCC is the total thickness of the ganglion cells and retinal nerve fibers	Change from baseline GCC thickness at 3 months
Pattern visual evoked potential (pVEP)	pVEP is an objective test that measures the electrical activity of the optical pathway in response to a light stimulus. The 120′ patterns reveals responses from the all retinal quadrants. The measurements were taken according to the ISCEV standards for both eyes. We used the 120'pattern VEP protocol, which combines p100 implicit time and amplitude, to create a numerical result	Change from baseline p100 latency and amplitude at 3 months

Sponsor
Ankara Universitesi Teknokent

Collaborators

- International Olympic Committee

Investigators
No information provided

General Publications

- Ozmert E, Arslan U. Management of retinitis pigmentosa by Wharton's jelly derived mesenchymal stem cells: preliminary clinical results. Stem Cell Res Ther. 2020 Jan 13;11(1):25. doi: https://doi.org/10.1186/s13287-020-1549-6.
- Ozmert E, Arslan U. Management of retinitis pigmentosa by Wharton's jelly-derived mesenchymal stem cells: prospective analysis of 1-year results. Stem Cell Res Ther. 2020 Aug 12;11(1):353. doi: https://doi.org/10.1186/s13287-020-01870-w.
- Arslan U, Ozmert E. Treatment of resistant chronic central serous chorioretinopathy via platelet-rich plasma with electromagnetic stimulation. Regen Med. 2020 Aug;15(8):2001-2014. https://doi.org/10.2217/rme-2020-0056. Epub 2020 Oct 27.
- Ozmert E, Arslan U. Management of Deep Retinal Capillary Ischemia by Electromagnetic Stimulation and Platelet-Rich Plasma: Preliminary Clinical Results. Adv Ther. 2019 Sep;36(9):2273-2286. https://doi.org/10.1007/s12325-019-01040-2. Epub 2019 Aug 5.
- Arslan U, Ozmert E. Management of Retinitis Pigmentosa via Platelet-Rich Plasma or Combination with Electromagnetic Stimulation: Retrospective Analysis of 1-Year Results. Adv Ther. 2020 May;37(5):2390-2412. https://doi.org/10.1007/s12325-020-01308-y. Epub 2020 Apr 18.

Anterior Segment

United States

Completed

Corneal Epithelial Stem Cells and Dry Eye Disease

ClinicalTrials.gov ID NCT03302273

Sponsor: Sloan W. Rush, MD
Information provided by Sloan W. Rush, MD, Panhandle Eye Group, LLP (Responsible Party)
Last Update Posted 2022-04-21

Brief Summary
To study of a novel, therapeutic Human Cells, Tissues, and Cellular and Tissue-Based Product (HCT/P) in the treatment of severe dry eye disease that is failing conventional treatments. This pilot study will carefully observe and monitor each qualifying and willing individual for response to treatment, signs of toxicity and adverse effects from the treatment, and for ability of the treatment to improve comfort and restore vision.

Detailed Description
Eligible patients presenting with severe dry eye disease that have failed treatment with conventional measures will be screened for inclusion into the study. The treatment will consist of administration of formulated topical eye drops containing cadaveric epithelial stem cell-derived product. These eye drops will be transplanted by topical self-administration by the patient four times daily in both eyes for a three-month interval.

Official Title
Topical Cadaveric-Derived Corneal Epithelial Stem Cell Transplant Therapy for the Treatment of Severe Dry Eye Disease

Conditions
Dry Eye Syndromes
Dry Eye
Ocular Inflammation
Ocular Surface Disease
Ocular Discomfort
Blepharitis

Study Type
Interventional

Enrollment (Actual)
17

Intervention/Treatment
Other: Corneal Epithelial Stem Cell Transplant

Phase
Not Applicable

Other Study ID Numbers
5

Study Start (Actual)
2019-02-01

Primary Completion (Actual)
2022-04-01

Study Completion (Actual)
2022-04-01

United States

Amarillo, Texas, United States, 79,109
Rush Eye Associates

Eligibility Criteria
Ages Eligible for Study
18 to 100 years (Adult, Older Adult)

Accepts Healthy Volunteer
No

Sexes Eligible for Study
All

Description

Inclusion Criteria
- Clinical diagnosis of advanced dry eye disease
- Severe dry eye symptoms despite having tried over a half dozen treatments

Exclusion Criteria
- Not willing to undergo clinical trial
- Unable to comply with treatment regimen

Primary Purpose: Treatment
Allocation: N/A
Interventional Model: Single Group Assignment
Masking: None (Open Label)

Arms and interventions

Participant group/arm	Intervention/treatment
Experimental: corneal epithelial stem cell transplant The treatment will consist of transplantation (via self-administration) of formulated topical eye drops containing cadaveric epithelial stem cell-derived biologic material four times daily in both eyes for a three-month interval	Other: Corneal Epithelial Stem Cell Transplant • Self-administration of formulated topical eye drops containing cadaveric epithelial stem cell-derived biologic material

Primary outcome measures

Outcome measure	Measure description	Time frame
Ocular surface disease index (OSDI)	Patient Reported Outcome by Survey	3 months

Secondary outcome measures

Outcome measure	Measure description	Time frame
Visual acuity	Snellen	3 months
Corneal topography	Automated corneal topographic indices	3 months

Sponsor
Sloan W. Rush, MD

Collaborators
No information provided

Investigators
No information provided

General Publications
No publications available

Belarus

Completed

Treatment of Patients With Inflammatory-dystrophic Diseases of the Cornea Using Autologous Stem Cells

ClinicalTrials.gov ID NCT04484402

Sponsor: Institute of Biophysics and Cell Engineering of National Academy of Sciences of Belarus
Information provided by Institute of Biophysics and Cell Engineering of National Academy of Sciences of Belarus (Responsible Party)
Last Update Posted 2020-07-30

Brief Summary
Treatment of patients with inflammatory-dystrophic diseases of the cornea using autologous limbal stem cells (corneal epithelial stem cells) or adipose-derived mesenchymal stem cells.

Detailed Description
In order to treat inflammatory and dystrophic diseases of the cornea, perilimbal injections of cultured autologous stem cells are performed. Corneal (limbal) epithelial stem cells (LSC) or mesenchymal stem cells derived from adipose tissue (ADSC) are mixed with sodium hyaluronate 1%.

Official Title
Treatment of Patients With Inflammatory-dystrophic Diseases of the Cornea Using Autologous Limbal Stem Cells (Corneal Epithelial Stem Cells) or Adipose-derived Mesenchymal Stem Cells

Conditions
Corneal Ulcer
Corneal Disease
Corneal Dystrophy

Study type
Interventional

Enrollment (Actual)
25

Intervention/Treatment
Biological: mesenchymal stem cells
Biological: limbal stem cells
Other: standard treatment

Phase
Phase 1 Phase 2

Other Study ID Numbers
IBCE_LSC

Study Start (Actual)
2016-10-03

Primary Completion (Actual)
2019-12-30

Study Completion (Actual)
2019-12-30

Belarus

Minsk, Belarus, 220,072
Institute of Biophysics and Cell Engineering of National Academy of Sciences of Belarus

Eligibility Criteria
Ages Eligible for Study
18 years and older (Adult, Older Adult)

Accepts Healthy Volunteer
No

Sexes Eligible for Study
All

Inclusion Criteria
- Inflammatory-dystrophic diseases of the cornea:
  - Epithelial defects of the cornea, ulcerative keratitis of various etiology, resistant to standard methods of treatment
  - Dystrophic diseases of the cornea, accompanied by edema
  - Burns of the cornea
  - Neurotrophic forms of keratitis
  - Persistent post-traumatic, postoperative, contusion keratitis, and keratopathy

Exclusion Criteria
- Pregnancy.
- Viral hepatitis B and C, HIV, tuberculosis, and other chronic viral and bacterial infections.
- Cancer or leukemia.

- Any diseases in the stage of decompensation.
- Mental disorders.
- Anomalies of eye refraction.

Primary Purpose: Treatment
Allocation: Non-Randomized
Interventional Model: Parallel Assignment
Masking: None (Open Label)

Arms and interventions

Participant group/arm	Intervention/treatment
Experimental: mesenchymal stem cells Patients with inflammatory-dystrophic diseases of the cornea receiving standard treatment plus adipose-derived mesenchymal stem cells	Biological: mesenchymal stem cells • Autologous adipose-derived mesenchymal stem cells mixed with sodium hyaluronate 1% solution Other: standard treatment • Standard treatment according to the Clinical protocols
Experimental: limbal stem cells Patients with inflammatory-dystrophic diseases of the cornea receiving standard treatment plus adipose-derived limbal stem cells	Biological: limbal stem cells • Autologous adipose-derived limbal stem cells mixed with sodium hyaluronate 1% solution Other: standard treatment • Standard treatment according to the Clinical protocols
Active Comparator: control Patients with inflammatory-dystrophic diseases of the cornea receiving standard treatment	Other: standard treatment • Standard treatment according to the Clinical protocols

Primary outcome measures

Outcome measure	Measure description	Time frame
Number of cured patients	Number of patients cured	2 month
Number of patients with treatment-related adverse events	MSC/LSC application-related adverse events assessed by blood count, liver function tests	4 weeks

Sponsor

Institute of Biophysics and Cell Engineering of National Academy of Sciences of Belarus

Collaborators

- Belarusian State Medical University

Investigators

- Study Director: Igor Volotovski, Prof, Head of the Lab of Institute of Biophysics and Cell Engineering
- Study Director: Liudmila Marchenko, Prof, Head of the Department of eyes diseases of Belarusian State Medical University
- Principal Investigator: Zinaida Kvacheva, Leading researcher, Institute of Biophysics and Cell Engineering

General Publications
No publications available

Results Overview
No Study Results Posted on ClinicalTrials.gov for this Study

Iran

Completed

The Improvement of Limbal Stem Cell Deficiency (LSCD) in Unilateral Stem Cell Damage by Amniotic Membrane Extract Eye Drop (AMEED)

ClinicalTrials.gov ID NCT02649621

Sponsor: Royan Institute
Information provided by Royan Institute (Responsible Party)
Last Update Posted 2016-01-07

Brief Summary
Corneal epithelial cells and limbal stem cells (LSC) are located in the limbus basal epithelium that are necessary for repair of corneal. About patients with deficient or absence of this area has been proposed various treatments such as limbal stem cell transplantation.

This study is a prospective clinical trial to compare the improvement of limbal stem cell deficiency (LSCD) in vivo by using of Amniotic Membrane Extract Eye Drop (AMEED).

Detailed Description
The cornea is the eye anterior portion which its refractive power and transparent is essential for normal vision.

The damaged cornea surface can reduce vision and leads to blindness, ultimately. Corneal epithelial and sometimes, limbal stem cells (LSC) are involved in corneal injuries. Limbal stem cells (LSC) are necessary for repair and reconstruction of corneal that reduction of these cells occurs by various causes including congenital causes, eye inflammatory diseases, and burns. For cornea stem cell damages or limbal stem cell deficiency (LSCD), cornea becomes conjunctivalization. LSCD may involve one eye (unilateral) or two eyes (bilateral).

Due to autologous transplantation problems in patients with unilateral LSCD and recent successes to resolve this problem, it seems that transplantation of cultured corneal stem cells on amniotic membrane is other way in treatment of unilateral LSCD. It is called ex vivo.

Amniotic membrane can modulate corneal epithelium healing by promoting re-epithelialization and migration of limbal stem cell while suppressing stromal inflammation, angiogenesis, and scarring. It is well accepted that amniotic membrane transplantation (AMT) as a temporary patch normally dissolves within 2 weeks. Consequent reapplication of membrane is difficult for the patient. On the other hand, in ex vivo, corneal tissue from healthy eye must be transported to laboratory for cell culture on AM that is the required equipment. There is, also, a risk of cell infection and transmission that is a very important issue. However, in infected cells, re-biopsy of the healthy eye is required that is uncomfortable and difficult for the patient. Other studies have been reported that amniotic membrane extract (AME) has same characteristics and features.

We previously have reported an effective potential of AMEED in limbal stem cell proliferation in vitro and also rabbit corneal epithelium healing in vivo.

This study is a prospective clinical trial to use Amniotic Membrane Extract Eye Drop (AMEED) on in vivo cultured limbal stem cells in the treatment of unilateral corneal stem cell damage.

Official Title
The Effect of in Vivo Cultured Limbal Stem Cells in the Treatment of Unilateral Corneal Stem Cell Damage

Conditions
Limbal Stem Cell Deficiency (LSCD)

Study Type
Interventional

Enrollment (Actual)
10

Intervention/Treatment
Biological: Amniotic Membrane Extract Eye Drop (AMEED)

Phase
Phase 1

Other Study ID Numbers
Royan-Eye-002

Study Start
2011-08

Primary Completion (Actual)
2012-03

Study Completion (Actual)
2012-08

Eligibility Criteria
Ages Eligible for Study
18 to 50 years (Adult)

Accepts Healthy Volunteer
No

Sexes Eligible for Study
All

Inclusion Criteria
- Patient with unilateral chemical burn (up to grade III); estimation of chemical burns amount and classification was according to Dua.
- Without Age limitation

Exclusion Criteria
- Lack of timely referral of patients for examinations
- Simultaneous use of other drugs that cause impairment of the data

Primary Purpose: Treatment
Allocation: N/A
Interventional Model: Single Group Assignment
Masking: None (Open Label)

Arms and interventions

Participant group/arm	Intervention/treatment
Experimental: Amniotic Membrane Extract Eye Drop recipient Patients with limbal stem cell Deficiency who receive amniotic membrane transplantation and amniotic membrane extract eye drop as eye drop	Biological: Amniotic Membrane Extract Eye Drop (AMEED) • Use of Amniotic Membrane Extract Eye Drop (AMEED) on in vivo cultured limbal stem cells in the treatment of unilateral corneal stem cell damage

Primary outcome measures

Outcome measure	Measure description	Time frame
Limbal defect size	The measurement of limbal defect size by microscope 12 months after corneal surgery	12 months
Corneal epithelial integrity	The measurement of Corneal epithelial integrity by microscope 12 months after corneal surgery	12 months
Corneal epithelial stability	The measurement of Corneal epithelial stability by microscope 12 months after corneal surgery	12 months

Secondary outcome measures

Outcome measure	Measure description	Time frame
Vascularization	Evaluation the rate of corneal vascularization with microscope 12 months after surgery	12 months
Transparency	Evaluation the transparency with visual tests 12 months after surgery	12 months
Best spectacle-corrected visual activity	Evaluation the best spectacle-corrected visual activity with visual tests	12 months

Sponsor
Royan Institute

Collaborators
No information provided

Investigators

- Study Chair: Hamid Gourabi, PhD, Head of Royan Institute
- Study Director: Marzieh Ebrahimi, PhD, Department of Regenerative Biomedicine at Cell Science Research Center, Royan Institute for Stem Cell Biology and Technology, ACECR, Tehran, Iran.
- Study Director: Alireza Baradaran, MD, Labafi Nejad Eye Research Center, Tehran, Iran
- Principal Investigator: Mitra Akbari, MD, Labafi Nejad Eye Research Center, Tehran, Iran
- Principal Investigator: Niloufar Shayan, MSc, Department of Stem Cells and Developmental Biology at Cell Science Research Center, Royan Institute for Stem Cell Biology and Technology, ACECR, Tehran, Iran.
- Principal Investigator: Maryam Momeni, MSc, Department of Stem Cells and Developmental Biology at Cell Science Research Center, Royan Institute for Stem Cell Biology and Technology, ACECR, Tehran, Iran.

General Publications
No publications available

Completed

Autologous Transplantation of Cultivated Limbal Stem Cells on Amniotic Membrane in Limbal Stem Cell Deficiency (LSD) Patients

ClinicalTrials.gov ID NCT00736307

Sponsor: Royan Institute
Information provided by Royan Institute
Last Update Posted 2011-07-15

Brief Summary
The purpose of this study is to evaluate the efficacy, safety, and long-term outcomes of ex vivo cultured limbal stem cell on amniotic membrane transplantation for corneal surface reconstruction in cases of partial and sever limbal stem cell deficiency.

Detailed Description
Ocular surface is composed of corneal and conjunctival epithelial cells. Normal cornea is essential for normal vision. Defects in renewal and repair of ocular surface as a result of limbal stem cell deficiency are now known to cause varying ocular surface morbidity including persistent photophobia, repeated and persistent surface breakdown, and overt conjunctivalization of the cornea. Restoring ocular health in these eyes has traditionally been frustrating. Ex vivo cultured limbal epithelial stem cells have been used successfully to treat limbal stem cell deficiency. Ex vivo limbal stem cell allograft transplantation is achieved by harvesting limbal corneal tissue from donor eyes (either matched living relatives or cadaveric donors). The donor stem cells are obtained by excising a small area of the conjunctiva at the limbus and

is a minor procedure. The tissue so obtained is then grown in tissue culture and once the cells have multiplied sufficiently, small sheets are transplanted on to the affected eye(s), backed with an amniotic membrane. The surgery is undertaken under either local or general anesthesia.

No Results Posted

Korea

Completed

Safety Evaluation of LSCD101 Transplantation for Limbal Stem Cell Deficiency

ClinicalTrials.gov ID NCT04773431

Sponsor: CliPS Co., Ltd.
Information provided by CliPS Co., Ltd. (Responsible Party)
Last Update Posted 2021-02-26

Brief Summary

Clinical Study Objective
To evaluate the tolerability and safety after transplantation of LSCD101 in patients with intractable limbal stem cell deficiency

Study Method
Subjects who finally meet the inclusion/exclusion criteria results of the subject eligibility evaluation will receive transplantation of the investigational product. Adverse drug reaction will be confirmed during 24 weeks after transplantation.

Official Title
Phase I Study to Evaluate Safety of LSCD101 (Cultured Autologous Limbal Epithelial Cell Sheet) Transplantation for Limbal Stem Cell Deficiency

Conditions
Limbus Cornea
Limbus Cornea Insufficiency Syndrome

Study Type
Interventional

Enrollment (Actual)
6

Intervention/Treatment
Drug: LSCD101 (Cultured Autologous Limbal Epithelial Cell Sheet) transplantation

Phase
Phase 1

Other Study ID Numbers
LSCD101_P1

Study Start (Actual)
2020-01-31

Primary Completion (Actual)
2021-01-14

Study Completion (Actual)
2021-01-14

Seoul, Korea, Republic of, 04168

Eligibility Criteria
Ages Eligible For Study
19 years and older (Adult, Older Adult)

Accepts Healthy Volunteer
No

Sexes Eligible For Study
All

Inclusion Criteria
- Men and women aged 19 or older
- Diagnosed with intractable limbal stem cell deficiency in single eye, with severity in which corneal neovascularization invades more than two quadrants
- No improvement of indication by known treatment (nonsurgical/surgical) method according to investigator's judgment
- BCVA 20/40 or less at screening in the eye with the limbal stem cell deficiency
- Voluntarily agreed to participate in this study and signed a consent form

Exclusion Criteria
- Inappropriate for harvesting limbal epithelial cells for the production of limbal epithelial cell sheet.
- Difficulty in reconstruction of the corneal epithelium because the eye is not closed due to eyelid function problems in the eye with the limbal stem cell deficiency.
- Severe dry eye with problem with tear secretion in both eyes (value of Schirmer I test without anesthesia <2 mm/5 min at screening)
- Acute ocular surface inflammation of both corneas at Visit 2
- Malignant tumor history (except in case of no recurrence more than 5 years after surgery)
- Uncontrolled comorbidities such as moderate to severe infections and bleeding
- Positive for virus infection (HBV, HCV, HIV, CMV, HTLVI/II, syphilis)

- Uncontrolled diabetes (HbA1c ≥ 9.0%)
- Uncontrolled hypertension (SBP > 140 mmHg or DBP > 90 mmHg despite taking antihypertensives at screening)
- Uncontrolled glaucoma (IOP > 21 mmHg despite taking glaucoma medication)
- Pregnant or lactating women
- Women with childbearing potential, those who do not agree to contraception by medically accepted contraceptive methods during clinical trials
- Participation in another clinical study within 4 weeks
- Subjects judged by the investigator to be inappropriate to participate in the clinical study

Primary Purpose: Treatment
Allocation: N/A
Interventional Model: Single Group Assignment
Masking: None (Open Label)

Arms and interventions

Participant group/arm	Intervention/treatment
Experimental: LSCD101 (Cultured Autologous Limbal Epithelial Cell Sheet) transplantation LSCD101 (Cultured Autologous Limbal Epithelial Cell Sheet) is transplanted in the lesion	Drug: LSCD101 (Cultured Autologous Limbal Epithelial Cell Sheet) transplantation • LSCD101 (Cultured Autologous Limbal Epithelial Cell Sheet) is transplanted in the lesion

Primary outcome measures

Outcome measure	Measure description	Time frame
Rate of adverse event	Adverse event	From limbal tissue collection until 24 weeks after transplantation

Secondary outcome measures

Outcome measure	Measure description	Time frame
Rate of normal/abnormal results in clinical laboratory test	Comparisons between pre/posttransplantation are summarized and presented	Before treatment, 1 week, 2 weeks, 4 weeks, 12 weeks, 24 weeks after transplantation
Blood pressure(systolic/diastolic) at each visit	Comparisons between pre/posttransplantation are summarized and presented	Before treatment, 1 day, 1 week, 2 weeks, 4 weeks, 12 weeks, 24 weeks after transplantation
Pulse rate at each visit	Comparisons between pre/posttransplantation are summarized and presented	Before treatment, 1 day, 1 week, 2 weeks, 4 weeks, 12 weeks, 24 weeks after transplantation
Rate of normal/abnormal results in Electrocardiogram	Comparisons between pre/posttransplantation are summarized and presented	Before treatment, 4 weeks, 12 weeks, 24 weeks after transplantation

Rate of normal/abnormal results in physical examination at each visit	Comparisons between pre/posttransplantation are summarized and presented	Before treatment, 4 weeks, 12 weeks, 24 weeks after transplantation
Rate of eye infection and inflammation	Eye infection and inflammation test (slit lamp microscopy)	Before treatment, 1 week, 2 weeks, 4 weeks, 12 weeks, 24 weeks after transplantation
Intraocular pressure test	Intraocular pressure test	Before treatment, 4 weeks, 12 weeks, 24 weeks after transplantation

Sponsor
CliPS Co., Ltd.

Collaborators
No information provided

Investigators
- Principal Investigator: SoHyang Chung, MD, PhD, The Catholic University of Korea

General Publications
No publications available

Spain

Completed

Safety Study of Stem Cell Transplant to Treat Limbus Insufficiency Syndrome

ClinicalTrials.gov ID NCT01562002

Sponsor: Instituto Universitario de Oftalmobiología Aplicada (Institute of Applied Ophthalmobiology)—IOBA
Information provided by Instituto Universitario de Oftalmobiología Aplicada (Institute of Applied Ophthalmobiology)—IOBA (Responsible Party)
Last Update Posted 2015-01-08

Brief Summary
The purpose of the study is to determine whether allogenic bone marrow stem cell transplant is safe and effective in the treatment of limbus insufficiency syndrome versus allogenic limbus stem cell transplant.

Official Title
Advanced Therapy for Ocular Surface Reconstruction. Allogenic Limbus Epithelial Stem-cell Transplant vs Bone Marrow Mesenchymal Stem-cell Transplant in Limbus Insufficiency Syndrome. Double-masked Randomized Trial

Conditions
Limbus Cornea Insufficiency Syndrome

Study Type
Interventional

Enrollment (Actual)
27

Intervention/Treatment
Procedure: Stem Cell with Amniotic Membrane Transplant

Phase
Phase 1 Phase 2

Other Study ID Numbers
IOBA-05-2010

Study Start
2012-03

Primary Completion (Actual)
2014-12

Study Completion (Actual)
Valladolid, Spain, 47,011
IOBA

Eligibility Criteria
Ages Eligible for Study
18 years and older (Adult, Older Adult)

Accepts Healthy Volunteer
No

Sexes Eligible for Study
All

Inclusion Criteria
- Male or Female older than 18 years.
- Signed Informed consent.
- Negative pregnancy test at inclusion for any potential childbearing female.
- Compromise of contraceptive method during all trial for any potential childbearing female.
- Diagnosis of Ocular Surface Failure due to Limbus Insufficiency Syndrome, based in any of the published characteristics as corneal surface neovascularization, loss of corneal transparency, epithelial irregularities, history of punctate keratitis, erosions or repetitive ulcers, and presence of symptoms and confirmed by the presence of epithelial phenotype cells assessed with conjunctival impression cytology.
- Availability for all the scheduled visits during the study.

Exclusion Criteria

- Uncontrolled systemic disease (e.g., hypertension or diabetes) or any disease that under medical decision might put the patient at risk during the surgery or follow-up examinations or may cause any hazard in data analysis.
- Active ocular infection in any eye. If the infection can be cured, inclusion can be considered after 30 days of inactive infection since its end.
- Alterations in lid statics/dynamics or any other pathology (e.g., severe dry eye syndrome) except the one that originated the Limbus insufficiency that under medical opinion might alter the results. Any of these must be corrected 3 months prior to patient inclusion, before reconsidering rescreening.
- Limbus insufficiency syndrome which has not been previously treated with all medical (not surgical) procedures available.
- No availability for all scheduled visits during the study.
- Any other circumstance under investigator's opinion that prevents patient inclusion even though normal inclusion and exclusion criteria are met.

Primary Purpose: Treatment
Allocation: Randomized
Interventional Model: Parallel Assignment
Masking: Double (Participant, Investigator)

Arms and Interventions

Participant group/arm	Intervention/treatment
Experimental: Bone Marrow mesenchymal stem cell Allogenic Bone Marrow mesenchymal stem cell in amniotic membrane transplant	Procedure: Stem Cell with Amniotic Membrane Transplant • Single stem cell expansion in amniotic membrane transplant
Active Comparator: Allogenic limbal stem cell Transplant Stem Cell with Amniotic Membrane Transplant	Procedure: Stem Cell with Amniotic Membrane Transplant • Single stem cell expansion in amniotic membrane transplant

Primary outcome measures

Outcome measure	Measure description	Time frame
Viability and safety of mesenchymal stem cell transplant	Absence of cell metaplasia with phenotype different to corneal or conjunctival as expected, confirmed by corneal impression cytology and in vivo confocal laser ophthalmoscopy	1 year

Secondary outcome measures

Outcome measure	Measure description	Time frame
Absence of complications in pre and peri surgical implantation	Correct handling and implantation of stem cell with amniotic membrane transplant following an 4-step classification	1 week
Improvement of two lines in Best-Corrected Visual Acuity	Improvement in visual acuity compared to baseline values before transplant	12 month

Sponsor
Instituto Universitario de Oftalmobiología Aplicada (Institute of Applied Ophthalmobiology)—IOBA

Collaborators
- University of Valladolid
- Centro en Red de Medicina Regenerativa de Castilla y Leon
- Red de Terapia Celular

Investigators
- Principal Investigator:Margarita Calonge, MD, PhD, Instituto Universitario de Oftalmobiología Aplicada (Institute of Applied Ophthalmobiology)—IOBA

General Publications
No publications available

Completed

Autologous Adipose-Derived Adult Stem Cell Implantation for Corneal Diseases (ADASCs-CT-CD) (ADASCs-CT-CD)

ClinicalTrials.gov ID NCT05279157

Sponsor: Vissum, Instituto Oftalmológico de Alicante
Information provided by Vissum, Instituto Oftalmológico de Alicante (Responsible Party)
Last Update Posted 2023-05-06

Brief Summary
Cellular therapy of the corneal stroma with implantation of mesenchymal stem cells derived from autologous adipose tissue with or without a carrier (scaffold) composed by decellularized human donor corneal stroma is used in patients with corneal diseases such as corneal dystrophies, and keratoconus.

For this purpose, the study planned to assess the enhancement of visual acuity, pachymetric, and aberrometric parameters with implantation of autologous mesenchymal adipose tissue-derived adult stem cells (ADASCs) alone, 120 μm thickness of decellularized or recellularized laminas with ADASCs. Three groups will be included in the study: (1) Implantation of a single dose of ADASCs alone without scaffold. (2) Implantation of decellularized human corneal lamina without ADASCs. (3) Implantation of the recellularized human corneal lamina with ADASCs.

Detailed Description
Different types of stem cells have been used in various ways in several research projects to find the optimal procedure to regenerate the human corneal stroma. It included several approaches which can be classified as intrastromal implantation of

stem cells (1) alone, (2) together with a biodegradable scaffold, (3) with a nonbiodegradable scaffold, or (4) with a decellularized corneal stromal scaffold. The complex structure of the corneal stroma has not been yet replicated, and there are well-known drawbacks to the use of synthetic scaffold-based designs.

Recently, several corneal decellularization techniques have been described, which provide an acellular corneal extracellular matrix (ECM). These scaffolds have gained attention in the last few years. The scaffold provides a more natural environment for the growth and differentiation of cells when compared with synthetic scaffolds. In addition, components of the ECM are generally conserved among species and are tolerated well even by xenogeneic recipients.

Keratocytes are essential for remodeling the corneal stroma and for normal epithelial physiology. This highlights the importance of transplanting a cellular substitute together with the structural support (acellular ECM) to undertake these critical functions in corneal homeostasis.

To the best of the investigators' knowledge, all attempts to repopulate decellularized corneal scaffolds have used corneal cells, but these cells have major drawbacks that preclude their autologous use in clinical practice (damage of the donor tissue, lack of cells, and inefficient cell subcultures), thus the efforts to find an extraocular source of autologous cells. A recent study by the investigators has shown a perfect bio integration of human decellularized corneal stromal laminas (100 μm thickness) with and without h-ADASC colonization inside the rabbit cornea in vivo, without observing any rejection response despite the graft being xenogeneic. The investigators also demonstrated the differentiation of h-ADASCs into functional keratocytes inside these implants in vivo, which then achieved their proper biofunctionalization. According to the investigator's opinion, the transplant of stem cells together with decellularized corneal ECM would be the best technique to effectively restore the thickness of a diseased human cornea, like in keratoconus. Through this technique, and using extraocular mesenchymal stem cells from patients, it is possible to transform allergenic grafts into functional autologous grafts, theoretically avoiding the risk of rejection.

The process flow is defined as following: (1) the file study: which start by receiving the file of the patient, the file will be forwarded to appointed physician's coordinators for review and submission of medical report, then the medical report will be evaluated within the cell therapy committee and the patient will be asked for clinical examination, and after consultation a reply to the patient with medical decision will follow with an approval or not to be recruited and if yes, a brief report about the procedure will be submitted and explained in details to the patient, a consent form must be signed if the patient agree to be included in the study, (2) the patient admission: which may start by completion of the procedure forms and doing the pre-op evaluation (initial workup defined as a clinical and biological assessment upon C.A.S which may include unaided and best spectacle-corrected visual acuity, refraction, slit lamp examination, intraocular pressure, fundoscopy, corneal topography, aberrometry, endothelial cell count (specular microscopy), corneal confocal microscopy, as well as blood tests), all these tests should be effectuated by an specialist and reviewed by an ophthalmologist. Then, a lipoaspiration of the subcutaneous adipose tissue to be performed by a plastic surgeon, sample processing at a cGMP facility for isolation, and characterization of the stromal vascular fraction enriched with ADASCs as well as laminas preparing (the quality control assessment will be realized before and after all the steps of the procedure starting from the ADASC collection to implantation including cell culture, cell quiescence, decellularization of human corneal lamina, recellularization of the laminas with ADASCs, and implantation; the assessment will include cell viability, cell number, cell apoptosis, immunophenotyping, infection, inflammation by analyzing the secreted cytokines from ADASCs, lamina cutting, lamina thickness, immunostaining, confocal microscopy, etc.); (3) the delivery (implantation) which starts by a peribulbar or retrobulbar anesthesia where the patient is placed under operating microscope, followed by a femtosecond laser-assisted mid-stromal lamellar dissection, then the autologous ADASCs or laminas or recellularized laminas will be placed within the pocket, and finally closed by a superior incision closure.

The patient will be put under antibiotic/steroids for a defined period and followed by the team at 1 week, 2 weeks, 1 month, 3 months, 6 months.

This protocol of cell therapy will be applied exclusively at VISSUM (Alicante-Spain), Barraquer Ophthalmology Center (Barcelona-Spain), Ramon y Cajal Hospital (Madrid-Spain), Asturias Prince Hospital (Madrid-Spain), and Murcia Hospital (Murcia-Spain), in affiliation with Miguel Hernandez of Elche University (Elche-Spain).

The patient's recruitment will take place: for the lipoaspiration, processing, preparing the cellularized laminas, and quality control assessments in VISSUM (Alicante-Spain), Barraquer Ophthalmology Center (Barcelona-Spain), Ramon y Cajal Hospital (Madrid-Spain), and Murcia Hospital (Murcia-Spain). All these steps are managed by VISSUM (Prof. J. ALIO).

Official Title
Implantation of Autologous Mesenchymal Stem Cells of Adipose Origin for the Treatment of Corneal Diseases

Conditions
Ophthalmological Disorder
Corneal Dystrophy
Treatment
Therapy
Keratoconus

Study Type
Interventional

Enrollment (Actual)
15

Intervention/Treatment
Procedure: Implantation
Procedure: Lipoaspiration

Phase
Phase 2

Other Study ID Numbers
EI17-00523
2018-000523-14 (EudraCT Number)

Study Start (Actual)
2022-04-19

Primary Completion (Actual)
2023-02-05

Study Completion (Actual)
2023-02-05

Alicante, Spain, 03016
Jorge L. Alio

Eligibility Criteria
Ages Eligible for Study
18 to 60 years (Adult)

Accepts Healthy Volunteer
No

Sexes Eligible for Study
All

Description

Inclusion Criteria
- Patients affected by corneal stromal dystrophies of any type, but particularly keratoconus, showing clear evidence in the ophthalmic examination of the presence and clear expression of the disease and loss of vision as a result of it.
- Patients with best-corrected visual acuity less than 0.6.
- Absence of chronic or recurrent inflammation in the anterior segment and on the ocular surface.
- Patient suitable to undergo corneal graft surgery under local anesthesia, from a medical point of view.
- Presurgical analysis of serum biochemistry and normal hematology.
- Serology for Human Immunodeficiency Virus (HIV), Hepatitis B (HBV), and Hepatitis C (HCV) negative.
- No history of malignancy.
- Normal chest x-ray (Rx).
- Normal urinalysis.
- Normal thyroid exam.

Exclusion Criteria

- Dense and extensive corneal stromal scars with severe involvement of the visual axis and located in the pupillary area, causing a decrease in the best-corrected vision to levels of 0.1 or less.
- Distance-corrected vision with glasses of 0.7 or greater.
- Extreme corneal thinning with risk of perforation.
- Infection.
- Previous corneal surgeries.
- Moderate or severe dry eye.
- Moderate or severe chronic inflammatory pathology of the ocular surface.
- Previous eye surgery other than cataract.
- Presence of cataract or other severe opacity of the transparent media of the eye that could prevent adequate examination of the fundus.
- Other ophthalmic comorbidity such as glaucoma or uveitis or any that requires the chronic use of topical ocular medication.
- Known and severe coagulation abnormalities.
- Any medical condition that may interfere with causing serious adverse effects during the study.
- Presence of active or inactive corneal neovascularization (CNV) in the eye to be treated.
- Any immunodeficiency or systemic autoimmune disease.
- Any current or intermittent immunosuppressive therapy or low-dose corticosteroids.
- Renal insufficiency, defined by creatine value>1.3 mg/dl.
- Serological evidence of hepatitis B, hepatitis C, or HIV infection.
- Pregnant or lactating woman.
- Corrected visual acuity in the eye contralateral to the experimental eye less than 20/40 (0.5).

Design Details

Primary Purpose: Treatment

Allocation: Randomized

Interventional Model: Parallel Assignment

Interventional Model Description: In this study, 15 evaluable patients affected by corneal stromal dystrophies of any type (particularly keratoconus) will be included in three different groups. Ten patients will be included in the two experimental groups and five in the control group with cell-free treatment (five patients per group and a total of three groups).

Masking: Triple (Participant Care Provider Investigator).

Masking Description: Patients who agree to participate in the study and sign the informed consent will be randomized to enter one of the three groups in a 1:1:1 ratio. For this, a randomization scheme will be used that will be unknown to the patient, the center staff, and the promoter staff who intervene directly in the study.

Arms and interventions

Participant group/arm	Intervention/treatment
Experimental: ADASCs Group Five patients, who will receive adipose tissue mesenchymal stem cells in a single dose as study treatment	Procedure: Implantation • Procedure: Implantation. The ophthalmologist will carry out the implantation of the ADASCs cells with or without scaffold in the intrastromal cornea of the patient • Other Names: – corneal implantation Procedure: Lipoaspiration • Procedure: Lipoaspiration. The patient will have a liposuction surgery from which the autologous ADASCs implant in the cornea will be obtained. Procedure: Implantation. The ophthalmologist will carry out the implantation of the ADASCs with or without scaffold in the intrastromal cornea of the patient • Other Names: – corneal implantation
Experimental: Acellular laminas group Five patients, who will receive decellularized corneal laminas as treatment medication	Procedure: Implantation • Procedure: Implantation. The ophthalmologist will carry out the implantation of the ADASCs cells with or without scaffold in the intrastromal cornea of the patient • Other Names: – corneal implantation
Experimental: ADASCs recellularized laminas group Five patients, who will receive adipose tissue mesenchymal cells combined with decellularized corneal laminas as study treatment in a single intervention	Procedure: Implantation • Procedure: Implantation. The ophthalmologist will carry out the implantation of the ADASCs cells with or without scaffold in the intrastromal cornea of the patient • Other Names: – corneal implantation Procedure: Lipoaspiration • Procedure: Lipoaspiration. The patient will have a liposuction surgery from which the autologous ADASCs implant in the cornea will be obtained. Procedure: Implantation. The ophthalmologist will carry out the implantation of the ADASCs with or without scaffold in the intrastromal cornea of the patient • Other Names: – corneal implantation

Primary outcome measures

Outcome measure	Measure description	Time frame
Improvement in best-corrected distance visual acuity (BCDVA)	Best-corrected visual acuity (BCDVA) will be studied in Snellen charts by Logmar and the corresponding equivalent in decimal scale	at 12 months

Secondary outcome measures

Outcome measure	Measure description	Time frame
Increase in corneal thickness	Corneal thickness will be evaluated by ultrasonic and optical pachymetry (Scheimpflug corneal topography)	at 12 months
Refractive changes	Refraction will be studied by refractive subjection performed by a certified good clinical practice technician	at 12 months
Improvement in anterior corneal surface regularity	Anterior corneal surface regularity will be evaluated by corneal topography map and analyzed by the changes in the Zernike polynomials (third-order and fourth-order spherical aberration)	at 12 months

Sponsor
Vissum, Instituto Oftalmológico de Alicante

Collaborators
- Universidad Miguel Hernandez de Elche
- Instituto de Salud Carlos III

Investigators
- Principal Investigator: Jorge L Alio, MD, PhD, Universidad Miguel Hernandez, Elche, Spain

General Publications
No publications available

Results Overview
No Study Results Posted on ClinicalTrials.gov for this Study

Chapter 4
Unknown Status Ocular Stem Cell Studies

Belgium

Limbal Epithelial Stem Cell Transplantation: a Phase II Multicenter Trial (MLEC)

ClinicalTrials.gov ID NCT02318485

Sponsor Ethisch Comité UZA
Information provided by Ethisch Comité UZA, University Hospital, Antwerp (Responsible Party)
Last Update Posted 2016-08-31

Study Overview

Brief Summary
This project aims to conduct a multicenter human clinical trial to effectively diagnose and treat limbal stem cell deficiency (LSCD) which can be caused by a number of different disease processes, all of which have a common denominator, i.e., a loss of the normal functioning epithelial stem cells of the cornea. A number of different diseases such as aniridia, Stevens–Johnson's syndrome, ocular cicatricial pemphigoid, bacterial keratitis, or chemical and thermal injuries to the cornea can lead to gradual loss of its functioning epithelial stem cells. This leads to vascularization of the cornea with surrounding conjunctival epithelium growing over it resulting in its scarring and opacification. These corneas, since vascular, no longer retain their immune privilege and prove extremely challenging to the ophthalmic surgeon. Grafting donor corneas which in normal circumstances (nonvascular corneas) would give excellent results, exhibit high rates of rejection. This can be very frustrating for both the patient as well as the surgeon. Since there are a number of different contributing disease pathologies, our estimate is that there are approximately 100 patients per year in Belgium that

J. N. Weiss, *Ocular Stem Cell Studies*,
https://doi.org/10.1007/978-3-031-58557-9_4

suffer from some degree of LSCD. These patients are very often not correctly diagnosed since this is a relatively new disease first described in 1990 after the discovery of limbal epithelial stem cells in 1989. The low rates of diagnosis could potentially be increased by in vivo confocal scanning microscopy as a noninvasive screening test to detect early signs of LSCD and therefore offer conservative treatment options. In the cases where total limbal stem cell deficiency has already developed, conventional corneal grafting is not suitable due to the high vascularity of the host bed. In this situation, we propose transplanting standardized limbal epithelial stem cell grafts generated from the patient's own ocular stem cells (from the good eye, where available) or from HLA-matched donors. Tiny 1 by 2 mm superficial limbal (peripheral cornea) biopsy can be taken and the epithelial stem cells expanded in the laboratory on a biological substrate, until there is a graft measuring >8 mm in diameter generated in a 2-week period. This can then be transplanted onto the diseased eye after removal of scar tissue using a novel surgical technique we have developed.

Detailed Description

This project aims to conduct a multicenter, single-arm phase II study, in which we will assess the safety and efficacy of transplanting standardized, non-xenogenic limbal stem cell grafts for patients with LSCD. The Good Manufacturing Practice (GMP) clean room facilities at the Centre for Cell Therapy and Regenerative Medicine (CCRG, UZA) will conduct tissue engineering, while the Departments of Ophthalmology at the University Hospitals in Antwerp (UZA), Brussels (UZB), and Gent (UZG) will enrol patients, perform preoperative tests, harvest the limbal biopsies, perform the auto- or allogenic graft transplantation and follow-up. Based on the past experience from our phase I/II mono-center human clinical trial, we have been able to document the safety and feasibility of the transplantation. 12/18 patients (2/3 allogenic and 10/15 autologous) have shown decrease in corneal vascularization, complete epithelialization, and incorporation of the graft into the anterior cornea post limbal stem cell transplantation. However, our current results are limited with respect to the sample size, feasibility of allogenic transplantation (only 3 cases), long-term follow-up (mean of 22 months, range 4–43 months), and disease monitoring strategies. Longer patient follow-up to determine long-term success (especially following subsequent corneal transplants) and inclusion of a larger allogenic patient population need to be addressed before this treatment can be translated to the clinics and recognized by the RIZIV as a reimbursable cell therapy.

It is pertinent to say that this cell-based therapy has the potential to provide this orphan pathology with an effective treatment option. The Flemish society will benefit from this project by improved rehabilitation of patients suffering from corneal blindness due to LSCD. By providing this treatment option, we can offer a ray of hope to patients that would formerly have to learn to live with their disability. By restoring vision for these patients, it is possible to reintegrate them into society as socially accepted individuals. Even a small degree of visual rehabilitation can make all the difference. For example, a slight increase in visual acuity can restore a person's ability to drive rather than having them rely exclusively on public transportation. More independence and a better quality of life are the main objectives that we are striving to achieve as the major outcome of this project.

Official Title
Translational Stem Cell Research in Ophthalmology—Regenerating the Anterior Cornea Through Standardized Transplantation of Limbal Epithelial Stem Cells: a Phase II Multicenter Trial

Conditions
Limbal Stem Cell Deficiency

Study Type
Interventional

Enrollment (Estimated)
60

Intervention/Treatment
Biological: Limbal epithelial stem cell graft

Phase
Phase 2

Other Study ID Numbers
CCRG13-003

Study Start
2015-04

Primary Completion (Estimated)
2019-12

Study Completion (Estimated)
2020-12

Belgium

Brussels, Belgium, 1090
Recruiting
University Hospital Brussels

Contact
Josephine Behaegel, MD
+32/2/477 Josephine.Behaegel@uzbrussel.be

Principal Investigator
Marcel Ten Tuscher, MD

Gent, Belgium, 9000
Recruiting
University Hospital, Ghent

Contact
Josephine Behaegel, MD
Josephine.Behaegel@uza.be

Principal Investigator
Bart P Leroy, MD

Antwerp Locations

Edegem, Antwerp, Belgium, 2650
Recruiting
Antwerp University Hospital

Contact
Josephine Behaegel, MD
+3238214524 Josephine.Behaegel@uza.be

Contact
Eline Melsbah
Eline.Melsbach@uza.be

Sub-Investigator
Sorcha Ni Dhubhghaill, MD/PhD Show more

Eligibility Criteria
Ages Eligible for Study
18 Years and older (Adult, Older Adult)

Accepts Healthy Volunteer
No

Sexes Eligible for Study
All

Inclusion Criteria
- Adults ≥18 years
- Voluntary written informed consent
- Patients suffering from LSCD IIa and IIb. Those suffering from IIc may be included once inflammation has subsided and cornea can be staged as IIb.
- Absence of any psychological, familial, sociological, or geographical condition potentially hampering compliance with the study protocol and follow-up schedule; those conditions should be discussed with the patient before registration in the trial.
- Women of childbearing potential should use adequate contraception prior to study entry and for the duration of study participation. Negative serum or urine β-HCG pregnancy test at screening.

Exclusion Criteria
- Subjects who are pregnant or lactating
- Subjects who have sensitivity to drugs that provide local anesthesia
- Subjects suffering from active infection of the external eye
- Medical conditions that prohibit the use of systemic immunosuppression (in cases of allogenic transplantation)

Primary Purpose: Treatment
Allocation: N/A
Interventional Model: Single Group Assignment
Masking: None (Open Label)

Arms and interventions

Participant group/arm	Intervention/treatment
Experimental: transplant A limbal epithelial stem cell graft will be transplanted onto the limbal stem cell deficient eye after it has been debrided of fibrovascular pannus	Biological: limbal epithelial stem cell graft • Other names: – Limbo-amnion composite graft – Limbal stem cell graft

Primary outcome measures

Outcome measure	Measure description	Time frame
Visual acuity	Using Log MAR scale	1 year
Presence of persistent epithelial defects	Using fluorescein drops and cobalt blue filter photography at the slit lamp	3 months post-surgery
Presence of corneal conjunctivalization	Slit-lamp photography and in vivo confocal microscopy	1 year
Change in corneal vascularization	Slit-lamp photography	1 year
Pain	Using visual pain scale	1 year
Photophobia	Using a 4-point scale	1 year
Rejection	Increased neovascularization, epithelial defects, and a drop in visual acuity redness pain irritation slit-lamp examination	1 year

Sponsor
Ethisch Comité UZA

Collaborators
- Universitair Ziekenhuis Brussel
- University Hospital, Ghent

Investigators
No information provided

General Publications
No publications available

Unknown Status

Cultivated Stem Cell Transplantation for the Treatment of Limbal Stem Cell Deficiency (LECT)

ClinicalTrials.gov ID NCT00845117

Sponsor Ethisch Comité UZ Antwerpen
Information provided by Ethisch Comité UZ Antwerpen, University Hospital, Antwerp (Responsible Party)
Last Update Posted 2013-06-20

Brief Summary
The purpose of this study is to determine whether cultivated stem cell transplantation is effective for the treatment of patients with corneal stem cell deficiency.

Official Title
Cultivated Stem Cell Transplantation for the Treatment of Limbal Stem Cell Deficiency

Conditions
Limbal Stem Cell Deficiency

Study Type
Interventional

Enrollment (Actual)
21

Intervention/Treatment
Procedure: cultivated limbal stem cell graft transplantation

Phase
Phase 1 Phase 2

Other Study ID Numbers
EC 7/28/153

Study Start
2008-07

Primary Completion (Estimated)
2014-06

Study Completion (Estimated)
2014-06

Belgium

Antwerp, Edegem, Belgium, 2650
University Hospital Antwerp

Eligibility Criteria
Ages Eligible for Study
(Child, Adult, Older Adult)

Accepts Healthy Volunteer
No

Sexes Eligible for Study
All

Inclusion Criteria
- Patients suffering from LSCD IIa and IIb. Those suffering from IIc may be included once inflammation has subsided and cornea can be staged as IIb.
- Absence of any psychological, familial, sociological, or geographical condition potentially hampering compliance with the study protocol and follow-up schedule; those conditions should be discussed with the patient before registration in the trial.
- Women of childbearing potential should use adequate contraception prior to study entry and for the duration of study participation

Exclusion Criteria
- Subjects who are pregnant or lactating
- Subjects who have sensitivity to drugs that provide local anesthesia
- Subjects suffering from active infection of the external eye
- Medical conditions that prohibit the use of systemic immunosuppression (in cases of allogenic transplantation)

Primary Purpose: Treatment
Allocation: N/A
Interventional Model: Single Group Assignment
Masking: None (Open Label)

Arms and interventions

Participant group/arm	Intervention/treatment
Experimental: limbal stem cell transplant The cornea is debrided of all superficial fibrovascular tissue and the cultivated stem cell graft is glued onto the cornea	Procedure: cultivated limbal stem cell graft transplantation • A limbal biopsy taken from the contralateral good eye in cases of unilateral disease or from a living related or cadaveric donor in cases of bilateral disease. The limbal stem cells from the biopsy are cultivated until a sheet of cells measuring approximately 12 mm in diameter is obtained. This is then ready for transplantation onto the diseased eye

Primary outcome measures

Outcome measure	Measure description	Time frame
Visual acuity		Just before surgery and up to 1 years after surgery
Conjunctivalization		Just before surgery and up to 1 years after surgery
Vascularization		Just before surgery and up to 1 years after surgery
Epithelialization		Just before surgery and up to 1 years after surgery

Secondary outcome measures

Outcome measure	Measure description	Time frame
Prolonged subsequent corneal graft survival time		1 year post corneal transplant

Sponsor
Ethisch Comité UZ Antwerpen

Collaborators
- Fund for Scientific Research, Flanders, Belgium
- European Society of Cataract and Refractive Surgeons
- Agentschap voor Innovatie door Wetenschap en Technologie

Investigators
- Principal Investigator: Nadia Zakaria, MBBS, PhD, University Hospital, Antwerp
- Principal Investigator: Carina Koppen, MD, PhD, University Hospital, Antwerp
- Study Director: Marie J Tassignon, MD, PhD, University Hospital, Antwerp
- Study Chair: Zwi Berneman, MD, PhD, University Hospital, Antwerp

General Publications

- Zakaria N, Koppen C, Van Tendeloo V, Berneman Z, Hopkinson A, Tassignon MJ. Standardized limbal epithelial stem cell graft generation and transplantation. Tissue Eng Part C Methods. 2010 Oct;16(5):921–7. https://doi.org/10.1089/ten.TEC.2009.0634.

Brazil

Unknown Status

Effect of Intravitreal Bone Marrow Stem Cells on Ischemic Retinopathy (RetinaCell)

ClinicalTrials.gov ID NCT01518842

Sponsor University of Sao Paulo
Information provided by Rubens Camargo Siqueira, University of Sao Paulo (Responsible Party)
Last Update Posted 2018-10-16

Brief Summary
This study aims to evaluate the behavior of the intravitreal use of bone marrow-derived stem cells in patients with ischemic retinopathy.

Detailed Description
Evaluate the effects on visual acuity, electroretinography, fluorescein angiography, and optical coherence tomography in 30 patients with ischemic retinopathy, including diabetic retinopathy with severe loss of retinal capillaries undergoing intravitreal injection of bone marrow-derived stem cells.

Official Title
Effect of Intravitreal Bone Marrow Stem Cells on Ischemic Retinopathy

Conditions
Ischemia

Study Type
Interventional

Enrollment (Estimated)
30

Intervention/Treatment
Biological: Intravitreal Bone Marrow Stem Cells

Phase
Not Applicable

Other Study ID Numbers
Siqueira Research Center

Study Start
2011-09

Primary Completion (Actual)
2018-10

Study Completion (Estimated)
2020-01

Brazil

Sao Jose do Rio Preto, SP, Brazil, 15010-100
Rubens Siqueira Research Center

Eligibility Criteria
Ages Eligible for Study
18 Years to 70 Years (Adult, Older Adult)

Accepts Healthy Volunteer
No

Sexes Eligible for Study
All

Description

Inclusion Criteria
- Diagnosis of Ischemic Retinopathy
- Enlargement of foveal avascular zone-FAZ ischemia or capillary drop out of >30% on Fluorescein retinography
- Logarithm of minimum angle of resolution (logMAR) BCVA of 1.0 (Snellen equivalent, 20/200) or worse
- Able (in the Investigator's opinion) and willing to comply with all study requirements

Exclusion Criteria
- Previous ocular surgery other than cataract
- Presence of cataract or other media opacity that would influence ocular fundus documentation and adequate ERG and visual field evaluation
- Other ophthalmic disease like glaucoma and uveitis
- Previous history of blood disorders like leukemia
- Known allergy to fluorescein or indocyanine green

Design Details
Primary Purpose: Treatment
Allocation: N/A
Interventional Model: Single Group Assignment
Interventional Model Description: intravitreal injection of autologous bone marrow stem cells
Masking: None (Open Label)

Arms and interventions

Participant group/arm	Intervention/treatment
Experimental: test group intravitreal stem cell Open-label study of Ischemic Retinopathy patients with best-corrected visual acuity (BCVA) worse than 20/200 Intervention: Biological: intravitreal injection of autologous bone marrow stem cells	Biological: intravitreal bone marrow stem cells • One intravitreal injection of a 0.1-mL cell suspension containing around 10×10^6 bone marrow mononuclear stem cells (BMMSC). All treatments were performed by a single retinal specialist using topical proparacaine drops under sterile conditions (eyelid speculum and povidone-iodine). Autologous BMMSC were injected into the vitreous cavity using a 27 gauge needle inserted through the inferotemporal pars plana 3.0–3.5 mm posterior to the limbus. After the injection, central retinal artery perfusion was confirmed with indirect ophthalmoscopy. Patients were instructed to instill one drop of 0.3% ciprofloxacin into the injected eye four times daily for 1 week after the procedure

Primary outcome measures

Outcome measure	Measure description	Time frame
Change in size of FAZ at 48 weeks		1 day to 48 weeks

Secondary outcome measures

Outcome measure	Measure description	Time frame
Change in central foveal thickness and best-corrected visual acuity at 48 weeks		1 day to 48 weeks

Sponsor
University of Sao Paulo

Collaborators
No information provided

Investigators
- Principal Investigator: Rubens C Siqueira, MD, PhD, Rubens Siqueira Research Center

General Publications
- Siqueira RC, Messias A, Gurgel VP, Simoes BP, Scott IU, Jorge R. Improvement of ischaemic macular oedema after intravitreal injection of autologous bone marrow-derived haematopoietic stem cells. Acta Ophthalmol. 2015 Mar;93(2):e174–6. https://doi.org/10.1111/aos.12473. Epub 2014 Jun 22. No abstract available.

China

Unknown Status

MSC-Exos Promote Healing of MHs (MSCs)

ClinicalTrials.gov ID NCT03437759

Sponsor Tianjin Medical University
Information provided by Xiaomin Zhang, Tianjin Medical University Eye Hospital (Responsible Party)
Last Update Posted 2021-04-06

Brief Summary
Purpose: To assess the safety and efficacy of mesenchymal stem cells (MSCs) and MSC-derived exosomes (MSC-Exos) for promoting healing of large and refractory macular holes (MHs).

Hypothesis: MSC and MSC-Exo therapy may promote functional and anatomic recovery from MH. MSC-Exo therapy may be a useful and safe method for improving visual outcomes of surgery for refractory MHs.

Detailed Description
Based on the purpose and hypothesis, the participants with large and longstanding idiopathic MHs underwent vitrectomy, internal limiting membrane peeling, MSC or MSC-Exo intravitreal injection, and heavy silicon oil, air, 20% SF6, or 14% C3F8 tamponade. MSCs were isolated from human umbilical cord, and MSC-Exos were isolated from supernatants of MSCs via sequential ultracentrifugation. At the time of study enrollment, as well as physical examinations, best-corrected visual acuity (BCVA) and intraocular pressure will be measured and fundoscopy be performed. All diagnoses of MH are going to confirmed via spectral-domain optical coherence tomography (OCT), and the minimum linear diameter (MLD) of each MH will be measured parallel to the retinal pigment epithelium. The participants are going to be followed up for at least 6 months via BCVA measurement, fundoscopy, OCT, and physical examinations.

Official Title
Mesenchymal Stem Cells-Derived Exosomes Promote Healing of Large and Refractory Macular Holes

Conditions
Macular Holes

Study Type
Interventional

Enrollment (Estimated)
44

Intervention/Treatment
Biological: exosomes derived from mesenchymal stem cells (MSC-Exo)

Phase
Early Phase 1

Other Study ID Numbers
2015KY-05

Study Start (Actual)
2017-03-01

Primary Completion (Actual)
2018-05-30

Study Completion (Estimated)
2021-12-30

China

Tianjin, China
Tianjin Medical University Hospital

Eligibility Criteria
Ages Eligible for Study
Up to 80 Years (Child, Adult, Older Adult)

Accepts Healthy Volunteer
Yes

Sexes Eligible for Study
All

Inclusion Criteria
- Clinical diagnosis of idiopathic macular hole whose diameter is greater than 400 μm.

Exclusion Criteria
- Only one functional eye.
- In other clinical trials.
- Other diseases which can affect visual acuity, such as cataract, diabetic retinopathy, glaucoma, corneal diseases, etc.
- Eye had undergone vitrectomy or scleral buckling, cataract surgery, Nd:YAG laser less than 1 month ago.
- Contralateral eye has poor prognosis than the study eye.
- Idiopathic or autoimmune uveitis history.
- Aphakic eye.
- Physical condition is poor that cannot keep prone position.
- Secondary macular lesions.
- The equivalent spherical diopter of the study eye before any refractive correction or cataract surgery, which is greater than 6.0d or above 26 mm of the ocular axis of the study eye.
- Intraocular pressure is higher than 25 mmHg.
- Within ocular inflammation, such as eye blepharitis, scleritis, keratitis, and conjunctivitis.
- Systemic condition is poor, such as the poor control of diabetes and hypertension, myocardial infarction, cerebrovascular accident, renal failure, and so on, and the researchers assessed those who are unable to complete the trail.

Primary Purpose: Treatment
Allocation: Randomized
Interventional Model: Parallel Assignment
Masking: Single (Participant)

Arms and interventions

Participant group/arm	Intervention/treatment
Experimental: experimental group Our intervention is to add treatment of exosomes derived from mesenchymal stem cells (MSC-Exo) after pars plana vitrectomy (PPV) and ILM peeling	Biological: exosomes derived from mesenchymal stem cells (MSC-Exo) • After air–liquid exchange, 50 or 20 μg MSC-Exo in 10 μL PBS was dripped into vitreous cavity around MH, leaving 20% SF6 or air as tamponade
No intervention: control group Control group that receives treatment of only pars plana vitrectomy (PPV) and ILM peeling	

Primary outcome measures

Outcome measure	Measure description	Time frame
Macular holes closure	Minimum linear diameter (MLD) of the hole measured by OCT	Baseline to 24 weeks post-surgery

Secondary outcome measures

Outcome measure	Measure description	Time frame
Best-corrected visual acuity (BCVA)	BCVA using a Landolt C acuity chart method	Baseline to 24 weeks post-surgery

Sponsor
Tianjin Medical University

Collaborators
No information provided

Investigators

- Principal Investigator: Rong X Li, MD, PhD, Tianjin Medical University Eye Hospital

General Publications
No publications available

Unknown Status

Clinical Study of Subretinal Transplantation of Human Embryo Stem Cell-Derived Retinal Pigment Epitheliums in Treatment of Macular Degeneration Diseases

ClinicalTrials.gov ID NCT02749734

Sponsor Southwest Hospital, China
Information provided by ZhengQin Yin, Southwest Hospital, China (Responsible Party)
Last Update Posted 2018-01-31

Brief Summary
The purpose of this study was to determine the safety and therapeutic effect of subretinal transplantation of human embryo stem cell-derived retinal pigment epitheliums (hESC-RPE) in patients with macular degeneration diseases, and explore new treatment modalities for macular degeneration diseases (Age-related macular degeneration and Stargardt macular dystrophy).

Official Title
Clinical Study of Subretinal Transplantation of Human Embryo Stem Cell-Derived Retinal Pigment Epitheliums in Treatment of Macular Degeneration Diseases

Conditions
Macular Degeneration
Stargardt Macular Dystrophy

Study Type
Interventional

Enrollment (Estimated)
15

Intervention/Treatment
Procedure: Subretinal transplantation

Phase
Phase 1 Phase 2

Other Study ID Numbers
2013CB967002

Study Start
2015-05

Primary Completion (Estimated)
2018-12

Study Completion (Estimated)
2019-12

China

Chongqing, Chongqing, China, 400038
Southwest Hospital

Eligibility Criteria
Ages Eligible for Study
18 Years to 75 Years (Adult, Older Adult)

Accepts Healthy Volunteer
No

Sexes Eligible for Study
All

Inclusion Criteria

- Aging from 18 to 75 years
- Must have signed informed consent
- At least one visually impaired eye caused by macular degeneration diseases
- Cannot be effectively treated with conventional therapies
- Best-corrected visual acuity scores between 19 and 73 letter in ETDRS (early treatment diabetic retinopathy study) eye chart, including 19 and 73 (or the equivalent of Snellen eyesight from 20/400 to 20/40)
- Visual loss caused by macular degeneration diseases

Exclusion Criteria

- Eyes with concomitant diseases which will interfere the visual improvement of the study
- Active intraocular inflammation regardless of the grade of severity
- Active infection (e.g., conjunctivitis, keratitis, scleritis, uveitis, ophthalmia)
- History of uveitis
- Severe cataract, glaucoma, retinal blood vessels occlusion, retinal detachment, macular hole, vitreous-macula traction
- Iris neovascularization
- Patients who have only one functioning eye, or the best-corrected vision of untreated eye scores less than 24 letters in ETDRS chart (corresponding to 20/320 in Snellen chart)
- History of intraocular surgery
- Severe systemic diseases: Stroke, coronary heart disease, angina pectoris, renal insufficiency needing dialysis
- Allergic to sodium fluorescein
- Uncontrolled hypertension (systolic pressure >40 mmHg, or diastolic prcssure >90 mmHg)
- Coagulative function disorder
- System administration of drugs that are toxic to lens, retina, or optic nerve like hydroxychloroquine, phenothiazine, ethambutol, tamoxifen, etc.
- Involved in other clinical trials of any medicine within 1 month (or within 5 half-life periods)
- Have maternity plan in 6 months
- In pregnancy or lactation period.

Primary Purpose: Treatment
Allocation: N/A
Interventional Model: Single Group Assignment
Masking: None (Open Label)

Arms and interventions

Participant group/arm	Intervention/treatment
Experimental: hESC-RPE Subretinal transplantation of human embryo stem cell-derived retinal pigment epitheliums	Procedure: subretinal transplantation • Transplant hESC-RPE into subretinal space of patients with macular degeneration

Primary outcome measures

Outcome measure	Measure description	Time frame
Number of participants with treatment-related adverse events [safety and tolerability]	Patients with treatment-related adverse events caused by local rejection of implanted cells or systemic immunosuppression treatment	Up to 12 months

Secondary outcome measures

Outcome measure	Measure description	Time frame
Number of early treatment diabetic retinopathy study (ETDRS) letters participants can recognize	Visual acuity is reflected by number of ETDRS letters participants can recognize	Up to 12 months
Visual field as examined by static perimetry	Area and sensitivity of visual field are detected by static perimetry	Up to 12 months
Flash electroretinogram (FERG)	Retinal electrophysiological function is tested by FERG	Up to 12 months
Amplitude and latency of flash visual evoked potentials (FVEP)	Optic nerve function as assessed by FVEP	Up to 12 months
Multifocal electroretinogram (MFERG)	Local retinal function as assessed by MFERG	Up to 12 months

Sponsor
Southwest Hospital, China

Collaborators
No information provided

Investigators
No information provided

General Publications
No publications available

Unknown Status

Umbilical Cord Mesenchymal Stem Cells Injection for Ocular Corneal Burn

ClinicalTrials.gov ID NCT03237442

Sponsor Guangzhou Saliai Stem Cell Science and Technology Co. Ltd.
Information provided by Guangzhou Saliai Stem Cell Science and Technology Co. Ltd. (Responsible Party)
Last Update Posted 2017-08-02

Brief Summary
Ocular chemical burn is one of the causes of vision loss in China, and there is no satisfactory treatment. Human umbilical cord mesenchymal stem cells (UC-MSCs)

have the biological characteristics of self-renewal, immune regulation, multidirectional differentiation, and tissue repair. Our preliminary research showed that in corneal alkali injury rabbits, the UC-MSCs can accelerated the cornea repair, inhibited angiogenesis. The aim of this study is to access the efficacy and safety of UC-MSCs in the treatment of corneal burn in human.

Official Title
Placebo-Controlled, Randomized, Double-blind Trial of Umbilical Cord Mesenchymal Stem Cells Injection for Ocular Corneal Burn

Conditions
Ocular Corneal Burn

Study Type
Interventional

Enrollment (Estimated)
100

Intervention/Treatment
Biological: human umbilical cord mesenchymal stem cells
Biological: placebo

Phase
Phase 1 Phase 2

Other Study ID Numbers
SSCST-MSC-OCB-01

Study Start (Estimated)
2018-01-01

Primary Completion (Estimated)
2018-12-30

Study Completion (Estimated)
2019-06-30

Study Contact
Name: Wanling Chen
Phone Number: 0086-20-37792600
Email: chenwanling@saliai.com
China
Guangdong Locations

Guangzhou, Guangdong, China
The First Affiliated Hospital of Jinan University

Contact
Yanyan Ma

Eligibility Criteria
Ages Eligible for Study
18 Years to 60 Years (Adult)

Accepts Healthy Volunteer
No

Sexes Eligible for Study
All

Inclusion Criteria
- Must be ocular burns including chemically burned or the thermally burned;
- The subjects are willing to accept this research, and promise to coordinate with the researchers during the follow-up period;
- The subjects should abide by the laws and rules of the study.

Exclusion Criteria
- The visual acuity is blind in any of the eye;
- Have corneal perforation or have the corneal perforation tendency;
- Have been accepted surgery on eyeball after trauma;
- IOP ≥25 mmHg even after antiglaucoma;
- Have the history of other corneal disease or surgery;
- Have the history of radiotherapy or surgery in the eyeball;
- Associated with corneal ulcer or endophthalmitis;
- Uncontrolled hypertension (≥150/95 mmHg);
- Abnormal liver and renal function;
- The pregnant women.

Primary Purpose: Treatment
Allocation: Randomized
Interventional Model: Parallel Assignment
Masking: Double (Participant Investigator)

Arms and interventions

Participant group/arm	Intervention/treatment
Experimental: group 1	Biological: human umbilical cord mesenchymal stem cells • Human UC-MSCs: 0.2 mL (about $2 * 10^6$ cells) subconjunctival injection
Active comparator: group 2	Biological: placebo • Saline injection

Primary outcome measures

Outcome measure	Measure description	Time frame
The percent of cornea perforation		3 months

Guangzhou Saliai Stem Cell Science and Technology Co. Ltd.

Collaborators
No information provided

Investigators
No information provided

General Publications
No publications available

Unknown Status

Allogeneic Mesenchymal Stem Cells Transplantation for Primary Sjögren's Syndrome (pSS)

ClinicalTrials.gov ID NCT00953485

Sponsor The Affiliated Nanjing Drum Tower Hospital of Nanjing University Medical School
Information provided by The Affiliated Nanjing Drum Tower Hospital of Nanjing University Medical School
Last Update Posted 2009-08-20

Brief Summary
This study will explore safety and efficacy of allogeneic mesenchymal stem cells transplantation (MSCT) to treat patients with diagnosis of primary Sjögren's Syndrome (pSS) who have been resistant to multiple standard treatments. The underlying hypothesis is that the pSS condition is caused by an abnormal immune homeostasis that can be restored by MSCT.

Detailed Description
- To test a new approach using allogeneic-derived mesenchymal stem cell-based therapy (MSCT) to treat refractory primary Sjögren's Syndrome (pSS)
- To determine the disease-free survival in pSS patients treated with MSCT in terms of disease activity index, pSS serology, and salivary gland function
- To assess adverse events of allogeneic MSC transplantation
- To assess the association of disease activity index, pSS serology levels, and saliva flow rate at baseline with disease-free survival

Official Title
Clinical Trial of Allogeneic Mesenchymal Stem Cells Transplantation for Primary Sjögren's Syndrome—Phase I/II

Conditions
Sjogren's Syndrome
Mesenchymal Stem Cells

Study Type
Interventional

Enrollment (Estimated)
20

Intervention/Treatment
Biological: Allogeneic Mesenchymal Stem Cells (AlloMSC)

Phase
Phase 1 Phase 2

Other Study ID Numbers
NJGLYY002

Study Start
2009-06

Primary Completion (Estimated)
2010-12

Study Completion (Estimated)
2011-12

Study Contact
Name: Lingyun Sun, MD
Phone Number: +86-25-83105219
Email: lingyunsun2001@yahoo.com.cn

China

Nanjing, Jiangsu, China, 210008
Recruiting
the Affiliated Drum Tower Hospital of Nanjing University Medical School

Eligibility Criteria
Ages Eligible for Study
15 Years to 70 Years (Child, Adult, Older Adult)

Accepts Healthy Volunteer
No

Sexes Eligible for Study
All

Inclusion Criteria
- All patients fulfilled the international classification criteria (2002) for primary Sjögren's Syndrome, man or woman aged from 15 to 70 years old, Sjögren's Syndrome Disease Activity Index (SSDAI) ≥8.
- Stimulated whole saliva flow rate less than 1–6 mL/6 min.
- Patients must be informed of the investigational nature of this study and give written informed consent in accordance with the institutional and hospital guidelines.
- Women of reproductive potential may not participate unless they have agreed to use an effective contraceptive method while on this trial and for 12 months fol-

lowing treatment. Women of childbearing potential must have a pregnancy test performed within 72 h prior to initiation of treatment.

Exclusion Criteria

- End-stage renal failure.
- Severe cardiopulmonary compromise or other system failure.
- Active, uncontrolled infections.
- Pregnant or nursing women may not participate due to the possibility of fetal harm or harm to nursing infants from this treatment regimen.

Design Details

Primary Purpose: Treatment
Allocation: Non-Randomized
Interventional Model: Single Group Assignment
Masking: None (Open Label)

Arms and interventions

Intervention/treatment
Biological: allogeneic mesenchymal stem cells (AlloMSC) • Allogeneic mesenchymal stem cells will be infused intravenously (single dose, 10^6 cells/kg body weight)

Primary outcome measures

Outcome measure	Measure description	Time frame
Sjögren's syndrome disease activity index		Monthly

Secondary outcome measures

Outcome measure	Measure description	Time frame
pSS Serology (ANA, dsDNA, SS-A, SS-B)		Monthly
Improvement of salivary gland function (measured as stimulated saliva flow rate)		Monthly

Sponsor

The Affiliated Nanjing Drum Tower Hospital of Nanjing University Medical School

Collaborators

No information provided

Investigators

- Principal Investigator: Lingyun Sun, MD, The Affiliated Nanjing Drum Tower Hospital of Nanjing University Medical School

General Publications

- Sun L, Akiyama K, Zhang H, Yamaza T, Hou Y, Zhao S, Xu T, Le A, Shi S. Mesenchymal stem cell transplantation reverses multiorgan dysfunction in systemic lupus erythematosus mice and humans. Stem Cells. 2009 Jun;27(6):1421–32. https://doi.org/10.1002/stem.68.

- Liang J, Zhang H, Hua B, Wang H, Wang J, Han Z, Sun L. Allogeneic mesenchymal stem cells transplantation in treatment of multiple sclerosis. Mult Scler. 2009 May;15(5):644–6. https://doi.org/10.1177/1352458509104590.
- Zhou K, Zhang H, Jin O, Feng X, Yao G, Hou Y, Sun L. Transplantation of human bone marrow mesenchymal stem cell ameliorates the autoimmune pathogenesis in MRL/lpr mice. Cell Mol Immunol. 2008 Dec;5(6):417–24. https://doi.org/10.1038/cmi.2008.52.
- Sun LY, Zhang HY, Feng XB, Hou YY, Lu LW, Fan LM. Abnormality of bone marrow-derived mesenchymal stem cells in patients with systemic lupus erythematosus. Lupus. 2007;16(2):121–8. https://doi.org/10.1177/0961203306075793.

Unknown Status

Safety and Efficacy of Umbilical Cord Mesenchymal Stem Cell Therapy for Patients with Progressive Multiple Sclerosis and Neuromyelitis Optica

ClinicalTrials.gov ID NCT01364246

Sponsor Shenzhen Beike Bio-Technology Co., Ltd.
Information provided by Shenzhen Beike Bio-Technology Co., Ltd. (Responsible Party)
Last Update Posted 2012-11-28

Brief Summary
Multiple sclerosis (MS) has been classically regarded as a chronic inflammatory autoimmune demyelinating disease of the central nervous system, along with a considerable pathological heterogeneity. Neuromyelitis optica (NMO) is a severe inflammatory, demyelinating disease, and its clinical characteristics include recurrent optic neuritis and longitudinally extensive transverse myelitis.

Current therapies provide only modest control of progressive Multiple Sclerosis and Neuromyelitis Optica. Stem cell therapy might open a light in the horizon in approaching to an efficient treatment in progressive MS and NMO. In this study, the safety and efficacy of Human Umbilical Cord Mesenchymal Stem Cells transplantation will be evaluated in patients with progressive MS and NMO.

Official Title
Phase I/II Study of Umbilical Cord Mesenchymal Stem Cell Therapy for Patients With Progressive Multiple Sclerosis and Neuromyelitis Optica

Conditions
Progressive Multiple Sclerosis
Neuromyelitis Optica

Study Type
Interventional

Enrollment (Estimated)
20

Intervention/Treatment
Biological: human umbilical cord mesenchymal stem cells

Phase
Phase 1 Phase 2

Other Study ID Numbers
BKCR-MS-1.0(2010)

Study Start
2010-01

Primary Completion (Estimated)
2013-03

Study Completion (Estimated)
2014-12

Study Contact
Name: Yun Yun Xu
Email: xuyun20042001@yahoo.com.cn

China
Jiangsu Locations

Nanjing, Jiangsu, China
Recruiting
Nanjing University Medical College Affiliated Drum Tower Hospital

Contact
Yun Xu
xuyun20042001@yahoo.com.cn

Eligibility Criteria
Ages Eligible for Study
18 Years to 60 Years (Adult)

Accepts Healthy Volunteer
No

Sexes Eligible for Study
All

Inclusion Criteria
- McDonald Diagnosis of Multiple Sclerosis or 2006 Diagnosis of neuromyelitis optica.
- Aged 16–65 years.

- Disease duration ≥2 years.
- Poor response to steroid therapy.
- Written informed consent and follow the clinic trail procedure.

Exclusion Criteria

- Cardiac insufficiency; Renal insufficiency; hepatic insufficiency; Total bilirubin higher than 1.5 times of upper limit of normal value; AST/ALT higher than 2.5 times of upper limit of normal value.
- Combined Pneumonia or other Severe systemic bacterial infection.
- Severe drug allergic history or anaphylaxis to two or more food or medicine.
- Intracranial hypertension dues to other brain lesions (e.g., Brain cancer).
- HIV+, TPPA +, patients diagnosed as HBV or HCV.
- Tumor Markers +.
- Severe psychotic patients, cognitive dysfunction, or cannot understand or sign the Consent Form.
- Coagulation disorders.
- Uncontrolled hypertension after treatment, blood pressure ≥180 mmHg/110 mmHg.
- Pregnancy.
- Enrollment in other trials in the last 3 months.
- Other criteria the investigator consider improper for inclusion.

Design Details

Primary Purpose: Treatment
Allocation: Non-Randomized
Interventional Model: Single Group Assignment
Masking: None (Open Label)

Arms and interventions

Participant group/arm	Intervention/treatment
Experimental: human umbilical cord mesenchymal stem cells transplantation Intervention group	Biological: human umbilical cord mesenchymal stem cells • Participants will be given hUC-MSCs transplantation

Primary outcome measures

Outcome measure	Measure description	Time frame
Score of Expanded Disability Status Scale (EDSS)		1 year after treatment

Secondary outcome measures

Outcome measure	Measure description	Time frame
Visual evoked potential (VEP)		1 year after treatment
Brainstem auditory evoked potential (BAEP)		1 year after treatment
Somatosensory evoked potential (SEP)		1 year after treatment
Brain magnetic resonance imaging (MRI) scan		1 year after treatment

Sponsor
Shenzhen Beike Bio-Technology Co., Ltd.

Collaborators
- The Affiliated Nanjing Drum Tower Hospital of Nanjing University Medical School
- The Affiliated Nanjing Brain Hospital of Nanjing University Medical School
- Nanjing University Medical College Affiliated Wuxi Second Hospital
- Xuzhou Medical University
- The Second Hospital of Nanjing Medical University

Investigators
No information provided

General Publications
No information provided

Unknown Status

Transplantation of Autologous Oral Mucosal Epithelial Stem Cell Sheet for Treating Limbal Stem Cell Deficiency Disease

ClinicalTrials.gov ID NCT03015779

Sponsor He Eye Hospital
Information provided by He Eye Hospital (Responsible Party)
Last Update Posted 2017-01-10

Brief Summary
The aim of this trial is to investigate the therapeutic effects of application autologous oral mucosal epithelial cell sheets in the treatment of limbal stem cell deficiency disease.

Detailed Description
A number of external factors can result in limbal stem cell deficiency (LSCD), causing pain, pathological corneal epithelium regeneration, inflammation reaction, and seriously visual loss. For patients with unilaterally damaged eye, transplantation of limbal tissue from the healthy eye can be used in the past years in the risk of inducing limbal stem cell damaged in the healthy eye.

While graft of human leukocyte antigen matched allogenic limbal stem cells require immune-suppressive medications, it can cause side effects in the body. Over the past 10 years, the clinical trials of application of cultured autologous oral mucosal epithelial cell sheet to treat LSCD have been emerged as a promising strategy in many countries. However, little study has been reported in China. This study investigates the therapeutic effects of application autologous oral mucosal epithelial cell sheets in the treatment of limbal stem cell deficiency disease in China.

Official Title
Transplantation of Autologous Oral Mucosal Epithelial Stem Cell Sheet for Treating Limbal Stem Cell Deficiency Disease

Conditions
Limbal Stem Cell Deficiency

Study Type
Interventional

Enrollment (Estimated)
7

Intervention/Treatment
Other: cultured autologous oral mucosal epithelial cell sheet

Phase
Phase 1 Phase 2

Other Study ID Numbers
20161029

Study Start
2017-01

Primary Completion (Estimated)
2017-12

Study Completion (Estimated)
2018-12

Eligibility Criteria

Ages Eligible for Study
18 Years and older (Adult, Older Adult)

Accepts Healthy Volunteer
No

Sexes Eligible for Study
All

Inclusion Criteria
- Male or female aged ≥18 years old
- Normal intraocular pressure
- Normal Eyelid
- No cataract and retinal disease
- No ulcers and tumor in the mouth

Exclusion Criteria
- Severe Systemic infection
- History of acute phase of ocular and oral inflammation

- History of hyper sensibility or allergy to antibiotics or serum
- Women who are or may be pregnant
- Patients with infectious diseases
- Patients with other complications

Primary Purpose: Treatment
Allocation: N/A
Interventional Model: Single Group Assignment
Masking: None (Open Label)

Arms and interventions

Participant group/arm	Intervention/treatment
Experimental: experimental group Cultured autologous oral mucosal epithelial cell sheet	Other: cultured autologous oral mucosal epithelial cell sheet • A prospective study, before and after self-control

Primary outcome measures

Outcome measure	Measure description	Time frame
The area of reepithelialization of corneal epithelium	The preoperative status of the eye will be recorded on the basis of clinical drawings and pre and postoperative photographs. These pictures will be sent to two corneal experts in order to evaluate the results Judgement criterion: the results will be tested with a slit lamp: Normal: no epithelial loss; mild: the area of epithelial loss <1/3; moderate: the area of epithelial loss >1/3 <2/3; severe: the area of epithelial loss >2/3	3 months

Secondary outcome measures

Outcome measure	Measure description	Time frame
Improvement of best correct visual acuity		1 year
Decrease of corneal neovascularization		1 year

Sponsor
He Eye Hospital

Collaborators
No information provided

Investigators
No information provided

General Publications
No publications available

Egypt

Unknown Status

Safety Study of Use of Autologous Bone Marrow-Derived Stem Cell in Treatment of Age-Related Macular Degeneration

ClinicalTrials.gov ID NCT02016508

Sponsor Al-Azhar University
Information provided by Abdelhakim Mohamed Safwat, Al-Azhar University (Responsible Party)
Last Update Posted 2013-12-24

Brief Summary
The purpose of this Phase I/II study is to investigate the safety and preliminary efficacy of unilateral intravitreally transplantation of adult bone marrow stem cells in subjects with geographic atrophy secondary to age-related macular degeneration.

Detailed Description
This study is an open-label investigation of the safety and preliminary efficacy of unilateral intravitreal injection of autologous bone marrow stem cells in subjects with Geographic Atrophy secondary to Age-Related Macular Degeneration (AMD). Subjects will be enrolled based on specific inclusion/exclusion criteria and evaluated at regular posttransplant intervals.

Human central nervous system-stem cells will be transplanted by a vitreoretinal surgeon. The transplantation will be conducted in the eye with the inferior best-corrected visual acuity (BCVA) (i.e., the Study Eye). Only the Study Eye will undergo transplantation. The adult bone marrow stem cells will be administered into the vitreous cavity through a standard surgical approach.

Moxifloxacin 300 mg (once per day) will be administered orally to all subjects for a period of 5 days (2 days pre and postoperatively).

Subjects will be monitored frequently for a total of 1 year after adult bone marrow stem cells cell injection.

Official Title
Intravitreal Injection of Human Bone Marrow-Derived Mesenchymal Stem Cell in Patients With Dry Age-related Macular Degeneration (AMD)

Conditions
Age-Related Macular Degeneration

Study Type
Interventional

Enrollment (Estimated)
1

Intervention/Treatment
Drug: autologous bone marrow-derived stem cells

Phase
Phase 1 Phase 2

Other Study ID Numbers
NCT 4619792013
4619792013 (Other Grant/Funding Number) (OTHER_GRANT: Al-Azhar university)

Study Start
2013-03

Primary Completion (Estimated)
2015-04

Study Completion (Estimated)
2015-06

Study Contact
Name: Abdelhakim Mohamed Safwat, M.D.
Phone Number: +201005151919
Email: abdelhakimsafwat@gmail.com

Egypt

Cairo, Nasr City, Egypt
Recruiting
Al-Azhar university medical school (Benin-cairo) ophthalmology department

Principal Investigator
Abdelhakim Mohamed Safwat, MSc

Eligibility Criteria
Ages Eligible for Study
50 Years and older (Adult, Older Adult)

Accepts Healthy Volunteer
No

Sexes Eligible for Study
All

Inclusion Criteria
- Diagnosis of age-related macular degeneration with geographic atrophy (GA)
- Only patients with a specific degree and extent of GA will be eligible
- Subjects with best-corrected visual acuity (BCVA) of less than or equal to 20/400 in the Study Eye
- No prior or current choroidal neovascularization in either eye
- Able to provide written informed consent prior to any study-related procedures
- Agree to comply in good faith with all conditions of the study and to attend all required study visits

Exclusion Criteria

- Prior vitreal or retinal surgery in the previous 6 months
- Glaucoma
- Atrophic macular disease of any other cause
- Diabetic retinopathy or diabetic macular edema in either eye
- Previous organ, tissue, or bone marrow transplantation
- Autoimmune disease
- Allergy to moxifloxacin
- Current or prior malignancy (or is on chemotherapy)

Design Details

Primary Purpose: Treatment
Allocation: N/A
Interventional Model: Single Group Assignment
Masking: None (Open Label)

Arms and interventions

Participant group/arm	Intervention/treatment
Experimental: autologous bone marrow stem cells Use of autologous bone marrow-derived stem cells as intravitreal injection in AMD patients	Drug: autologous bone marrow-derived stem cells • Intravitreal injection of autologous bone marrow stem cells in 0.1 mL volume

Primary outcome measures

Outcome measure	Measure description	Time frame
Number of subjects with adverse events	Descriptive analysis of frequency and types of adverse events experienced by each subject during the study period	6 months

Secondary outcome measures

Outcome measure	Measure description	Time frame
Assessment of visual function changes from the base line	Assessment will include change in the mean of BCVA, optical coherence tomography imaging, fluorescein angiography, slit-lamp examination with fundus photography, electroretinographic evidence (mfERG) showing enhanced activity in the location	6 months

Sponsor

Al-Azhar University

Collaborators

- Hassan, Hosny, M.D. M.Sc
- Samour, Hany M.D. M.Sc
- Ismail, Mahmoud M.D. M.Sc
- Higazy, Hasan M.D. M.Sc
- Abou el kheir, Wael, M.D. M.Sc.
- Gabr, Hala, M.D. M.Sc.
- Bakry, Sayed, PhD. M.Sc.

Investigators
No information provided

General Publications
No publications available

India

Unknown Status

To Evaluate the Clinical Safety and Efficacy of Limbal Stem Cell for Treatment of Superficial Corneal Pathologies

ClinicalTrials.gov ID NCT04932629

Sponsor L.V. Prasad Eye Institute
Information provided by Vivek Singh, L.V. Prasad Eye Institute (Responsible Party)
Last Update Posted 2021-06-21

Brief Summary
This study proposes to investigate the transplantation of ex vivo cultivated allogenic limbal stromal cells for the treatment of the corneal pathologies. The limbus is an ideal source as the stem cells are numerous and located very superficially in the tissue. Preclinical work suggests human corneal stromal stem cells can be isolated from the cadaveric tissues, cultivated in conditions suitable for cell-based therapy and used to prevent fibrosis in a murine model of corneal stromal scarring. Further, these cells are able to successfully engraft, differentiate, and mediate wound healing in the corneal stroma such that the tissue remains healthy, free of fibrotic tissue, and optically transparent. The clinical implications of these findings are substantial in that it represents the potential to lessen the burden on donor tissue necessary for corneal allografts by using cultured cells to regenerate tissue. We foresee the ability of a clinician to and grow and expand the cells in number and after surgically removing the scar tissue from the wounded eye, apply the cultured limbal stem cells to regenerate healthy, transparent tissue.

Detailed Description
This would be a single-center prospective, open labeled, non-randomized interventional study. This study is an Investigator Initiated Study (IIS). The Ethics Committee of the LV Prasad Eye Institute, Hyderabad, would prospectively approve this study. This study would be conducted in strict adherence to the tenets of the Declaration of Helsinki, ICH GCP E6 (R2), Indian GCP Guidelines and New Drugs and Clinical Trial Rules 2019 and associated amendments and current National Ethical Guidelines for Biomedical and Health Research Involving Human Participants and Current National Stem Cell Research Guidelines and New Drugs and Clinical Trial Rules 2019.

Once the participants are found to be suitable for limbal transplant surgery, the patients will be administered written informed consent and audio/visual consent as per regulations. Detailed ophthalmic examination will be done to ensure that the patient is eligible for the trial.

All the screening procedures will be accomplished within 14 ± 7 days. On Day 0, that is the date of surgery, a Unique Participant Identification Number (UPIN) will be assigned to each patient and it would be in addition to hospital medical record number. The surgery will be done under local or general anesthesia (depending on age and patient preference).

In this prospective interventional study, patients with unilateral superficial corneal scars will undergo a surgical procedure. Limbal ring from a cadaveric donor tissue, which is therapeutically accepted and serologically tested, is collected. This tissue will then be cultivated in the stem cell biology laboratory using standardized culture technique. Briefly the limbal tissue will be cut up into small pieces and digested overnight using an enzyme (Collagenase L). The cells obtained from the digest will be cultured on a petri-dish using 2% serum and growth factors. The cultured cells will be passaged three times to remove all epithelial cells from the culture.

In the second procedure, the eligible patients will undergo corneal transplant surgery, when the central corneal epithelium will be removed using a surgical sponge. 0.1 mL of stromal cells in a concentration of 0.5×10^6 cells/μL diluted in the thrombin component of fibrin glue (TISEEL, Baxter) will be applied to the debrided corneal stroma. A soft bandage contact lens will be placed over the cornea at the end of the procedure. The patient will receive topical antibiotic and steroid eye drops in the postoperative period. Periodic comprehensive ophthalmic evaluation along with anterior segment optical coherence tomography (ASOCT) scanning and slit-lamp photography will be done at Day 1, Day 7, Day 30, Day 90, Day 180, and Day 360 and 720 Days post-surgery. The primary outcome measure of this study is to note any ocular or systemic adverse effects of this intervention at the various postoperative time points. The secondary outcome measures are visual improvement and change in the density and appearance of the corneal scarring and other pathologies after treatment.

Official Title

A Proof of Concept Study to Evaluate the Clinical Safety and Efficacy of Ex vivo Cultivated Allogenic Limbal Stem Cell Transplantation for Treatment of Superficial Corneal Pathologies

Conditions

Corneal Scars and Opacities

Study Type

Interventional

Enrollment (Estimated)

20

Intervention/Treatment

Biological: Ex vivo cultivated Allogeneic limbal stromal stem cells

Phase
Early Phase 1

Other Study ID Numbers
HERF-SSC-SCP-01

Study Start (Estimated)
2021-07

Primary Completion (Estimated)
2021-12

Study Completion (Estimated)
2023-01

Study Contact
Name: Dr. Sayan Basu, MBBS, MS
Phone Number: 040-68102534
Email: sayanbasu@lvpei.org

Study Contact Backup
Name: Dr. Vivek Singh, PhD
Phone Number: 040-68102286
Email: viveksingh@lvpei.org

Eligibility Criteria
Ages Eligible for Study
18 Years to 60 Years (Adult)

Accepts Healthy Volunteer
No

Sexes Eligible for Study
All

Inclusion Criteria
- Male and female participants who are ≥18 and ≤60 years of age
- Patients having unilateral superficial corneal pathologies (defined as involving the anterior 200 μM of the corneal stroma on ASOCT imaging)
- Corneal burns, ulcers and scars
- No prior history of corneal transplantation
- No ongoing and other active ocular pathology
- Candidate for stem cell transplant
- No severe pathological and psychological conditions that might interfere with the patient's participation in the study
- Able to provide written and audio-visual informed consent prior to any study specific screening procedures with the understanding that the patient has the right to withdraw from the study at any time, for any reason without prejudice

Exclusion Criteria

- Bilateral corneal disease.
- Corneal scars with limbal dysfunction (clinically defined as absent limbal palisades or conjunctivalization of the cornea).
- Ocular surface disease including dry eye disease (defined as a Schirmer's test of less than 10 mm at 5 min).
- Unknown etiology, postherpetic eye disease, or eyes with active intraocular inflammation.
- Children (<18 years of age).
- Less than 3 months after documented clinical resolution of acute disease.
- Inability/refusal to give written informed consent.
- Undergo any of the anterior segment imaging tests.
- Patient should have not participated in another clinical study within 30 days of their enrolment on this study.
- History or evidence of cardiac disease: congestive heart failure; New York Heart Association (NYHA) class 2 or greater; active coronary artery disease; unstable angina, cardiac arrhythmias requiring anti-arrhythmic therapy, atrio-ventricular block of second or third degree, or uncontrolled hypertension, patients with recent (less than 6 months) myocardial infarction (MI) or coronary revascularization.
- Pregnant and lactating patients, positive urine pregnancy test in women of child-bearing potential.
- Reproductive age patients not practicing effective and adequate birth control measures.
- Previous participation in this study.

Primary Purpose: Treatment
Allocation: N/A
Interventional Model: Single Group Assignment
Masking: None (Open Label)

Arms and interventions

Participant group/arm	Intervention/treatment
Experimental: experimental group The eligible patients will undergo corneal transplant surgery, when the central corneal epithelium will be removed using a surgical sponge. 0.1 mL of stromal cells in a concentration of 0.5×10^6 cells/μL diluted in the thrombin component of fibrin glue (TISEEL, Baxter) will be applied to the debrided corneal stroma	Biological: ex vivo cultivated allogeneic limbal stromal stem cells • In this prospective intervention study, patients with unilateral superficial corneal scars will undergo a surgical procedure and in the second procedure, the eligible patients will undergo corneal transplant surgery, when the central corneal epithelium will be removed using a surgical sponge. 0.1 mL of stromal cells in a concentration of 0.5×10^6 cells/μL diluted in the thrombin component of fibrin glue (TISEEL, Baxter) will be applied to the debrided corneal stroma. A soft bandage contact lens will be placed over the cornea at the end of the procedure. The patient will receive topical antibiotic and steroid eye drops in the postoperative period. Periodic comprehensive ophthalmic evaluation along with anterior segment optical coherence tomography (ASOCT) scanning and slit-lamp photography will be done at Day 1, Day 7, Day 30, Day 90, Day 180, and Day 360 and 720 Days post-surgery

Primary outcome measures

Outcome measure	Measure description	Time frame
Measurement of any ocular or systemic adverse effects	The primary outcome measure of this study is to note any ocular or systemic adverse effects	Day 90 post-surgery

Secondary outcome measures

Outcome measure	Measure description	Time frame
Measurement of visual improvement	Visual improvement using ETDRS vision chart method where visual acuity is the measurement, changes in ETDRS score is a scale and ETDRS letter score will be measured in log units	Day 720 post-surgery
Change in corneal light scattering	Change in corneal light scattering using Scheimpflug imaging where corneal light scattering is the measurement, corneal densitometry measurements is a scale and expressed in standardized grayscale units (GSU)	Day 720 post-surgery

Sponsor
L.V. Prasad Eye Institute

Collaborators
No information provided

Investigators
- Principal Investigator: Dr. Vivek Singh, PhD, LV Prasad Eye Institute (Hyderabad Eye Research Foundation)
- Principal Investigator: Dr. Sayan Basu, MBBS, MS, LV Prasad Eye Institute (Hyderabad Eye Research Foundation)

General Publications

- Basu S, Hertsenberg AJ, Funderburgh ML, Burrow MK, Mann MM, Du Y, Lathrop KL, Syed-Picard FN, Adams SM, Birk DE, Funderburgh JL. Human limbal biopsy-derived stromal stem cells prevent corneal scarring. Sci Transl Med. 2014 Dec 10;6(266):266ra172. https://doi.org/10.1126/scitranslmed.3009644.

Unknown Status

Stem Cells Therapy for Corneal Blindness (ExCell)

ClinicalTrials.gov ID NCT02948023

Sponsor L.V. Prasad Eye Institute
Information provided by Vivek Singh, L.V. Prasad Eye Institute (Responsible Party)
Last Update Posted 2018-03-06

Brief Summary
This is an investigative initiated, comparative pilot clinical trial to ascertain the safety of application of ex vivo cultivated limbal stem cells in human eyes for treating different superficial corneal pathologies. Preclinical work in murine models have already demonstrated efficacy of this technique in curing murine corneal pathologies.

Detailed Description
In this prospective interventional study, patients with superficial corneal pathologies like scars, ulcers, and burns will undergo a surgical procedure. Limbal ring from a cadaveric donor tissue, which is therapeutically accepted and serologically tested, is collected. This tissue will then be cultivated in the stem cell biology laboratory using standardized culture technique. Briefly the limbal tissue will be cut up into small pieces and digested overnight using an enzyme (Collagenase L). The cells obtained from the digest will be cultured on a petri-dish using 2% serum and growth factors. The cultured cells will be passaged three times to remove all epithelial cells from the culture. In the second procedure, the central corneal epithelium will be removed using a surgical sponge. 0.1 mL of stromal cells in a concentration of 5×10^3 cells/μL diluted in the thrombin component of fibrin glue (TISEEL, Baxter) will be applied to the debrided corneal stroma. A soft bandage contact lens will be placed over the cornea at the end of the procedure.

The patient will receive topical antibiotic and steroid eye drops in the postoperative period.

Periodic comprehensive ophthalmic evaluation along with anterior segment optical coherence tomography (ASOCT) scanning and slit-lamp photography will be done at day 1, day 7, day 45, and day 90.

Official Title
Ex vivo Cultivated Limbal Stem Cell Transplantation for Treatment of Superficial Corneal Pathologies

Conditions
Corneal Injuries
Corneal Burns
Corneal Scars and Opacities

Intervention/Treatment
Biological: Ex vivo cultivated limbal stem cell pool

Study Start (Actual)
2016-08-31

Primary Completion (Estimated)
2019-02

Study Completion (Estimated)
2019-08

Enrollment (Estimated)
100

Study Type
Interventional

Phase
Phase 1

Other Study ID Numbers
LVPEI-2014-1 Version 1 15 Jan 14

Study Contact
Name: Sayan Basu, MBBS, MS
Phone Number: 9140-30612625 ext 2625
Email: sayanbasu@lvpei.org

Study Contact Backup
Name: Vivek Singh, MSc PhD
Phone Number: 9140-30612286 ext 2286
Email: viveksingh@lvpei.org
India
Telangana Locations

Hyderabad, Telangana, India, 500034
Recruiting
LV Prasad Eye Institute

Contact
Sayan Basu, MBBS, MS
+9140 3061 2625 ext 2625 sayanbasu@lvpei.org

Contact
Vivek Singh, MSc PhD
+9140 3061 2286 ext 2286 viveksingh@lvpei.org

Eligibility Criteria
Inclusion Criteria
- Male and Female participants who are aged between 18 and 60 years.
- Patients who are having with unilateral blindness due to superficial (defined as involving the anterior 200 μm of the corneal stroma on ASOCT imaging) corneal wounds, ulcers, burns, and scars.
- Eyes with otherwise normal.

Exclusion Criteria
- Active infection, proven by microbiology.
- Unknown etiology, postherpetic eye disease, or eyes with active intraocular inflammation.
- Children (<18 years of age).
- Inability/refusal to give written informed consent or to undergo any of the anterior segment imaging tests.

Ages Eligible for Study
18 Years to 60 Years (Adult)

Sexes Eligible for Study
All

Accepts Healthy Volunteers
No

Design Details
Primary Purpose: Treatment
Allocation: Non-Randomized
Interventional Model: Parallel Assignment
Masking: None (Open Label)

Arms and interventions

Participant group/arm	Intervention/treatment
No intervention: standard surgical therapy Includes the control group that fulfills the inclusion criteria	
Active comparator: ex vivo cultivated limbal stem cell pool 0.5 million stromal and epithelial cells will be incorporated in 0.05 mL of commercially available fibrin glue and pasted over the corneal lesion after epithelial debridement	Biological: ex vivo cultivated limbal stem cell pool • 0.5 million stromal and epithelial cells will be incorporated in 0.05 mL of commercially available fibrin glue and pasted over the corneal lesion after epithelial debridement

Primary outcome measures

Outcome measure	Measure description	Time frame
Ocular or systemic adverse effects		Until 3 months after surgery

Secondary outcome measures

Outcome measure	Measure description	Time frame
Visual improvement after treatment	Checking for the improvement in visual acuity that will be calculated using LogMAR chart (where 20/20 vision is normal)	Until 3 months after surgery

Sponsor
L.V. Prasad Eye Institute

Collaborators
No information provided

Investigators
- Principal Investigator: Sayan Basu, MBBS, MS, LV Prasad Eye Institute
- Principal Investigator: Vivek Singh, MSc PhD, LV Prasad Eye Institute

General Publications
No publications available

Unknown Status

Clinical Study to Evaluate Safety and Efficacy of BMMNCs in Retinitis Pigmentosa

ClinicalTrials.gov ID NCT01914913

Sponsor Chaitanya Hospital, Pune
Information provided by Dr. Sachin Jamadar, Chaitanya Hospital, Pune (Responsible Party)
Last Update Posted 2014-09-17

Brief Summary
Retinitis pigmentosa is an eye disease in which there is damage to the retina. The retina is the layer of tissue at the back of the inner eye that converts light images to nerve signals and sends them to the brain. Autologous Bone Marrow-derived Mono Nuclear Stem Cells (BMMNCs) is used for this condition.

Detailed Description
This study will involve about 8 visits over 1½ years. Each participant will be enrolled in study as per inclusion and exclusion criteria. Informed consent will be taken from subject before including in study. Subject will undergo specific tests of the participant's vision and health.

These visits may include visual exams, blood draw for laboratory testing, brief medical history and exam, and occasionally a questionnaire (survey), in addition to the visit for the surgical procedures. The primary outcome for this study will be a visual acuity score after 1 year of stem cell therapy.

Follow-up visits will be required regularly to determine the effectiveness of Autologous Bone Marrow-derived Mono Nuclear Stem Cells (BMMNCs) in RP.

Official Title
An Open-Labeled Clinical Study to Evaluate the Safety and Efficacy of Autologous Bone Marrow-Derived Mono Nuclear Stem Cells (BMMNCs) in Retinitis Pigmentosa. It is Self-Funded (Patients' Own Funding) Clinical Trial

Conditions
Retinitis Pigmentosa
Intervention/Treatment
Biological: BMMNCs

Study Start
2014-09

Primary Completion (Estimated)
2016-11

Study Completion (Estimated)
2016-11

Enrollment (Estimated)
15

Study Type
Interventional

Phase
Phase 1 Phase 2

Other Study ID Numbers
CSCC/BMRP/2013//01

Study Contact
Name: Sachin P Jamadar, D.Ortho
Phone Number: 918888788880
Email: sac2751982@gmail.com
India
Maharashtra Locations

Pune, Maharashtra, India, 411030
Recruiting
Chaitanya Hospital

Contact
Sachin P Jamadar, D ORTHO
+918888788880 sac2751982@gmail.com

Principal Investigator
Anant E Bagul, MS, Ortho

Ages Eligible for Study
18 Years to 65 Years (Adult, Older Adult)

Sexes Eligible for Study
All

Accepts Healthy Volunteers
Yes

Eligibility Criteria
Inclusion Criteria
- Age 18–65
- Diagnosis of retinitis pigmentosa
- Willingness to undergo bone marrow and umbilical cord-derived mesenchymal stem cell transplantation.
- To give an informed consent as well as sign the required Informed Consent Form (ICF) for the study.
- Willingness to regularly visit the hospital/clinic for follow-up during the follow-up period/on prior agreed time points as per the protocol.

Exclusion Criteria

- Positive test results for HIV and AIDS complex, HCV (hepatitis C virus), HbsAg and Syphilis
- Women who are pregnant or lactating
- Complications of diabetic retinopathy.

Design Details

Primary Purpose: Treatment
Allocation: N/A
Interventional Model: Single Group Assignment
Masking: None (Open Label)

Arms and interventions

Participant group/arm	Intervention/treatment
Other: BMMNCs BMMNCs	Biological: BMMNCs • Intervention therapy, total 3 doses, in 30 days, in 7 days interval, transfer of autologous bone marrow-derived mono nuclear stem cell (BMMNCs)

Primary outcome measures

Outcome measure	Measure description	Time frame
Visual acuity	Electronic visual acuity (EVA) technology	1 year

Collaborators and Investigators

This is where you will find people and organizations involved with this study.

Sponsor

Chaitanya Hospital, Pune

Collaborators

No information provided

Investigators

- Principal Investigator: Anant E Bagul, MS Ortho, Chaitanya Hospital

General Publications

No publications available

Unknown Status

Study the Safety and Efficacy of Bone Marrow-Derived Autologous Cells for the Treatment of Optic Nerve Disease (OND)

ClinicalTrials.gov ID NCT01834079

Sponsor Chaitanya Hospital, Pune
Information provided by Dr. Sachin Jamadar, Chaitanya Hospital, Pune (Responsible Party)
Last Update Posted 2014-09-17

Brief Summary
This is a single-arm, single-center trial to check the safety and efficacy of bone marrow-derived autologous mono nuclear cell (100 million per dose) clinical trial to be conducted for 36 months in patients with optic nerve diseases.

Detailed Description
Optic nerve diseases: conditions which produce injury or dysfunction of the second cranial or optic nerve, which is generally considered a component of the central nervous system. Damage to optic nerve fibers may occur at or near their origin in the retina, at the optic disk, or in the nerve, optic chiasm, optic tract, or lateral geniculate nuclei. Clinical manifestations may include decreased visual acuity and contrast sensitivity, impaired color vision, and an afferent papillary defect. Primary output of clinical study is to measure improvement in visual loss and improvement in idiopathic intracranial. This is a single-arm, single-center trial to check the safety and efficacy of bone marrow-derived autologous mono nuclear cell (100 million per dose) clinical trial to be conducted for 36 months.

Official Title
Study the Safety and Efficacy of Bone Marrow-Derived Autologous Cells for the Treatment of Optic Nerve Disease. It is Self-Funded (Patients' Own Funding) Clinical Trial

Conditions
Optic Atrophy

Intervention/Treatment
Biological: Stem Cell Therapy

Study Start
2014-09

Primary Completion (Estimated)
2016-07

Study Completion (Estimated)
2016-07

Enrollment (Estimated)
24

Study Type
Interventional

Phase
Phase 1 Phase 2

Other Study ID Numbers
00104

Study Contact
Name: Sachin P Jamadar, D.Ortho
Phone Number: +918888788880
Email: sac2751982@gmail.com

Study Contact Backup
Name: Smita S Bhoyar, B.A.M.S. PGCR
Phone Number: 9372620569
Email: drsmitabhoyar@rediff.com
India
Maharashtra Locations

Pune, Maharashtra, India, 4
Recruiting
Chaitanya Hospital

Contact
Sachin P Jamadar, D. Ortho
8888788880 sac2751982@gmail.com

Contact
Smita S Bhoyar, BAMS.PGCR
9372620569 drsmitabhoyar@rediffmail.com

Principal Investigator
Anant E Bagul, MS

Ages Eligible for Study
18 Years to 50 Years (Adult)

Sexes Eligible for Study
All

Accepts Healthy Volunteers
Yes

Eligibility Criteria
Inclusion Criteria

- Patient Should suffer from Optic Nerve diseases like Diabetic Retinopathy, retinal pigmentation.
- Age in between 18 and 50.
- Willingness to undergo Bone Marrow-derived autologous cell therapy.
- Ability to comprehend the explained protocol and thereafter give an informed consent as well as sign the require Informed consent form (ICF) for the study.
- An ability and willingness to regular visit to hospital for protocol and follow-up.

Exclusion Criteria

- Patients with preexisting or Current systemic disease such as lung, liver (exception: History of uncomplicated Hepatitis A), gastrointestinal, Cardiac, Immunodeficiency (including HIV), or laboratory Investigation that could cause a neurological defect including Syphilis, clinically relevant polyneuropathy, etc.
- History of Life-threatening Allergic or immune-mediated reaction.
- Hemodynamically Unstable.

Design Details
Primary Purpose: Treatment
Allocation: N/A
Interventional Model: Single Group Assignment
Masking: None (Open Label)

Arms and interventions

Participant group/arm	Intervention/treatment
Other: stem cell intrathecal injection of MNC stem cell therapy	Biological: stem cell therapy • Autologous stem cells (MNCs) intrathecal transplantation of single-arm autologous MNCs 100 millions per dose in 3 divided doses at interval of 7 days, intrathecal/intravenous. Follow-up will be taken to 3 months or as required • Other names: – Intrathecal transplantation of autologous stem cells

Primary outcome measures

Outcome measure	Measure description	Time frame
Reduction in degeneration of the optic nerve with improvement in vision		6 months

Secondary outcome measures

Outcome measure	Measure description	Time frame
Increase in visual function		6 months
Improvement in idiopathic intracranial hypertension		6 months

Collaborators and Investigators
This is where you will find people and organizations involved with this study.

Sponsor
Chaitanya Hospital, Pune

Collaborators
No information provided

Investigators
- Principal Investigator: Anant E Bagul, M. S. Chaitanya Hospital

General Publications
No publications available

Unknown Status

Limbus-Derived Stem Cells for Prevention of Postoperative Corneal Haze

ClinicalTrials.gov ID NCT03295292

Sponsor L.V. Prasad Eye Institute
Information provided by Vivek Singh, L.V. Prasad Eye Institute (Responsible Party)
Last Update Posted 2019-02-26

Brief Summary
This is an investigator-initiated pilot clinical trial to ascertain the safety and efficacy of application of ex vivo cultivated limbal stem cells in human eyes for treating Corneal Haze after Photo-Therapeutic/Refractive Keratectomy (PTK/PRK) and Collagen Cross Linking (CXL). Instead of using adjunctive medical therapy like application of MMC (Mitomycin), a technique of cell delivery with fibrin sealant can be used. These cells are harvested from therapeutically accepted and serologically tested cadaveric corneas. The isolated limbal epithelial and mesenchymal or stromal cell suspension will then be cultured in CGMP laboratories and be tested for sterility. These cells have also been shown to be effective in treating haze in laser refractive surgery in an animal model. Our initial experience of using these cells in a previous clinical trial showed that they were effective in preventing corneal haze in patients with burns and ulcers.

Detailed Description
In this pilot clinical trial, patients undergoing Photo-Therapeutic/Refractive Keratectomy (PTK/PRK) and Collagen Cross Linking (CXL) will be given human Limbus-derived Corneal Stem cells to assess the safety of these cells. Cells will be cultivated in a cGMP laboratory using standardized culture technique; from the limbal rims of cadaveric corneoscleral donor tissues that are therapeutically accepted

and serologically tested. Briefly the limbal tissue will be cut up into small pieces and digested overnight using an enzyme (Collagenase L). The cells obtained from the digest will be cultured on a petri-dish using 2% serum and growth factors. A mixture of limbal epithelial cells and limbal stromal cells obtained from these cultures will be used in surgical procedures in a ratio of 2:1, after all the sterility checks. Mixed suspension of the limbal epithelial and the stromal cells at a concentration of 50,000 cells/μL diluted in the thrombin component of fibrin sealant (TISEEL, Baxter) will be applied. The primary outcome measure is safety of this treatment and the secondary outcome measure is their efficacy that will be assessed at 1 month time points by (1) Clinical photography using a standard method where multiple blinded observers will grade the clinically apparent change in haze and (2) Objective quantification of the amount of light scattering using densitometry analysis function in Oculus Pentacam (OCULUS Optikgeräte GmbH).

Official Title
Pilot Clinical Trial to Assess the Safety of Limbus-derived Corneal Stem Cells in Prevention of Corneal Haze After Photo-Therapeutic/Refractive Keratectomy (PTK/PRK) and Collagen Cross Linking (CXL)

Conditions
Corneal Scars and Opacities

Intervention/Treatment
Biological: Stem cells
Other: Vehicle

Study Start (Actual)
2017-08-01

Primary Completion (Estimated)
2019-12

Study Completion (Estimated)
2020-06

Enrollment (Estimated)
15

Study Type
Interventional

Phase
Phase 1

Other Study ID Numbers
LEC 07-17-068

Study Contact
Name: Sayan Basu, MBBS MS
Phone Number: +9140 3061 2625 ext 2625
Email: sayanbasu@lvpei.org

Study Contact Backup
Name: Pravin K Vaddavalli, MBBS MS
Phone Number: +91 403061 2626 ext 2626
Email: pravin@lvpei.org
India
Telangana Locations

Hyderabad, Telangana, India, 500034
Recruiting
LV Prasad Eye Institute

Contact
Sayan Basu, MBBS MS
+9140 3061 2625 ext 2625 sayanbasu@lvpei.org

Contact
Vivek Singh, MSc PhD
+9140 3061 2286 ext 2286 viveksingh@lvpei.org

Ages Eligible for Study
20 Years to 40 Years (Adult)

Sexes Eligible for Study
All

Accepts Healthy Volunteers
No

Eligibility Criteria
Inclusion Criteria
- Adult patients between 20 and 40 years of age
- Meeting standard selection criteria for bilateral PTK/PRK or CXL
- No systemic diseases
- Eligible to give informed consent
- No other ocular coexisting pathologies

Exclusion Criteria
- Undergoing surgery in only one eye
- Grossly asymmetric pathology
- Refusal to give informed consent
- Not agreeable or uncooperative for corneal imaging
- Unlikely to come for follow-up for 1 month
- International and outstation patients

Design Details
Primary Purpose: Treatment
Allocation: Randomized
Interventional Model: Parallel Assignment
Masking: Double (Participant Investigator)

Arms and interventions

Participant group/arm	Intervention/treatment
Sham comparator: standard surgery with vehicle After the standard surgical procedure (PRK/PTK or CXL), patients will receive 50 μL Fibrin Sealant (sham) without cells	Other: vehicle • 50 μL of commercially available fibrin sealant (Baxter, TISEEL)
Experimental: standard surgery with stem cells After the standard surgical procedure (PRK/PTK or CXL), patients will receive 50 μL of a mixture of limbus-derived corneal epithelial cells and limbus-derived corneal stromal cells in a ratio of 2:1, at a concentration of 50,000 cells/μL diluted in the thrombin component of Fibrin Sealant (TISEEL, Baxter)	Biological: stem cells • Mixture of limbus-derived corneal epithelial cells and limbus-derived corneal stromal cells in a ratio of 2:1, at a concentration of 50,000 cells/μL diluted in fibrin sealant Other: vehicle • 50 μL of commercially available fibrin sealant (Baxter, TISEEL)

Primary outcome measures

Outcome measure	Measure description	Time frame
Maintenance of preoperative best-spectacle-corrected visual acuity		1 month

Secondary outcome measures

Outcome measure	Measure description	Time frame
Efficacy in reducing corneal light scatter using Scheimpflug imaging	Efficacy in reducing corneal light scatter using Scheimpflug imaging at 1 month by: Objective quantification of the amount of light scattering using densitometry analysis function in Oculus Pentacam (OCULUS Optikgeräte GmbH) • All routine clinical and imaging protocols performed for patients undergoing PTK/PRK and CXL will be diligently followed	1 month

Sponsor
L.V. Prasad Eye Institute

Collaborators
No information provided

Investigators
- Principal Investigator: Sayan Basu, MBBS MS, LV Prasad Eye Institute
- Principal Investigator: Vivek Singh, MSc PhD, LV Prasad Eye Institute
- Principal Investigator: Jagadesh C Reddy, MBBS MD, LV Prasad Eye Institute
- Principal Investigator: Pratik Gogri, MBBS MS, LV Prasad Eye Institute

General Publications
No publications available

Jordan

Unknown Status

Treatment of Optic Neuropathies Using Autologous Bone Marrow-Derived Stem Cells

ClinicalTrials.gov ID NCT02638714

Sponsor Stem Cells Arabia
Information provided by Stem Cells Arabia (Responsible Party)
Last Update Posted 2020-03-17

Brief Summary
A single-arm, single-center trial to assess the safety and efficacy of restoring function in damaged optic nerves using autologous purified populations of bone marrow-derived stem cells (BM-SCs) through a 24 month follow-up period.

Detailed Description
Optic nerve atrophy (ONA) is a condition defined as a damage to the optic nerve that harmfully affects central and peripheral vision. ONA may occur as a result of optic neuritis, compression by tumors or aneurysms, toxic and nutritional neuropathies, trauma, or as a secondary complication to other systemic diseases such as diabetes. Symptoms of ONA vary diversely, but mainly include blurred vision and a reduction in optic sharpness and color visualization. ONA is irreversible process, and current medical strategies focus on finding the underlying cause, and trying to prevent further vision loss and protect the other healthy eye. This is a single-arm, single-center trial to assess the safety and efficacy of purified adult autologous bone marrow-derived CD34+, CD133+, and CD271+ stem cells through a 24 month follow-up period. The combination of these three cell types was based on their diverse potentialities to differentiate into specific functional cell types to regenerate damaged optic nerves and supporting issues and vasculature, and the availability of clinical-grade purification system (CliniMACS) and Microbeads to purify the target cell populations in clinically approved methods. Anticipated outcomes of this study are defined in an overall improvement of vision, restoration of functions to damaged optic nerves, and improvement in quality of life of patients.

Official Title
Treatment of Optic Neuropathies Using Autologous Bone Marrow-Derived Clusters of Differentiation (CD) 34+, 133+, and 271+ Stem Cells: A Pilot Study

Conditions
Optic Neuropathy

Intervention/Treatment
Biological: Stem Cells

Study Start
2013-04

Primary Completion (Estimated)
2021-09-01

Study Completion (Estimated)
2021-10-01

Enrollment (Estimated)
100

Study Type
Interventional

Phase
Phase 1 Phase 2

Other Study ID Numbers
SCA-ON1

Jordan

Amman, Jordan, 11953
Recruiting
Stem Cells of Arabia

Contact
Adeeb AlZoubi, PhD
adeebalzoubi@stemcellsarabia.net

Principal Investigator
Adeeb AlZoubi, PhD

Ages Eligible for Study
18 Years to 55 Years (Adult)

Sexes Eligible for Study
All

Accepts Healthy Volunteers
No

Eligibility Criteria
Inclusion Criteria

- Patient should suffer from Optic Nerve Atrophy diseases like diabetic retinopathy and retinal pigmentation.
- Age in between 18 and 55 years old.
- Willingness to undergo bone marrow-derived autologous cell therapy.
- Ability to comprehend the explained protocol.
- Ability and willingness to regularly visit to hospital for protocol and follow-up.

Exclusion Criteria

- Patients with preexisting or current systemic disease such as lung, liver, gastro-intestinal, cardiac, immunodeficiency, syphilis, or clinically relevant polyneuropathies.
- History of life-threatening allergic or immune-mediated reaction.

Ages Eligible for Study
18 Years to 55 Years (Adult)

Sexes Eligible for Study
All

Accepts Healthy Volunteers
No

Design Details
Primary Purpose: Treatment
Allocation: N/A
Interventional Model: Single Group Assignment
Masking: None (Open Label)

Arms and interventions

Participant group/arm	Intervention/treatment
Experimental: stem cells Intervention: transplantation of autologous purified stem cells	Biological: stem cells • Intervention: transplantation of autologous purified stem cells

Primary outcome measures

Outcome measure	Measure description	Time frame
Reduction in degeneration of the optic nerve using the visual field assessment with the Humphrey automated and Goldmann manual perimeters		24 months

Secondary outcome measures

Outcome measure	Measure description	Time frame
Improvement in visual function using the documentation of visual acuity using the Snellen chart		24 months

Sponsor
Stem Cells Arabia

Collaborators
No information provided

Investigators

- Principal Investigator: Adeeb Al Zoubi, PhD

General Publications
No publications available

Unknown Status

Autologous Bone Marrow-Derived CD34+, CD133+, and CD271+ Stem Cell Transplantation for Retinitis Pigmentosa

ClinicalTrials.gov ID NCT02709876

Sponsor Stem Cells Arabia
Information provided by Stem Cells Arabia (Responsible Party)
Last Update Posted 2020-03-17

Study Registration Dates

First Submitted
2016-02-26

First Submitted That Met QC Criteria
2016-03-10

First Posted (Estimated)
2016-03-16
Study Record Updates

Last Update Submitted That Met QC Criteria
2020-03-15

Last Update Posted
2020-03-17

Last Verified
2020-03

Brief Summary
A single-arm, single-center trial to evaluate the safety and efficacy of autologous purified populations of bone marrow-derived stem cells in patients with Retinitis Pigmentosa (BM-SCs) through a 48 month follow-up period.

Detailed Description
Retinitis pigmentosa (RP) is an inherited disorder of the photoreceptor cells in the retina. Patients may lose vision since they were young or later in life. Currently, there are more than 60 genes identified as the cause of this condition, one of which, RPE65, has been studied in several gene therapy trials for Leber congenital amaurosis with promising results. Another treatment approach for RP is stem cell therapy. Studies in animal models of RP have shown that subretinal injection of bone marrow-derived mesenchymal stem cells may delay degenerative changes of photoreceptor cells. This is a single-arm, single-center trial to assess the safety and efficacy of purified adult autologous bone marrow-derived CD34+, CD133+, and CD271+ stem cells through a 48 month follow-up period. The combination of these three cell types was based on their diverse potentialities to differentiate into specific

functional cell types to regenerate damaged retinal tissue, and the availability of clinical-grade purification system (CliniMACS) and microbeads to purify the target cell populations in clinically approved methods.

Conditions
Retinitis Pigmentosa

Intervention/Treatment
Biological: Stem Cell Transplantation

Study Start
2014-04

Primary Completion (Estimated)
2020-09

Study Completion (Estimated)
2021-03

Enrollment (Estimated)
50

Study Type
Interventional

Phase
Phase 1 Phase 2

Other Study ID Numbers
SCA-RP1

Jordan

Amman, Jordan, 11953
Stem Cells Arabia

Ages Eligible for Study
18 Years to 55 Years (Adult)

Sexes Eligible for Study
All

Accepts Healthy Volunteers
No

Eligibility Criteria
Description

Inclusion Criteria
- Retinitis pigmentosa patients diagnosed by ophthalmologists.
- Visual acuity (measured with ETDRS) less than or equal to 20/70 and visual field below 30° central in both eyes.
- Best-corrected visual acuity less than 6/120 by Snellen visual acuity chart.

Exclusion Criteria

- Other eye conditions that could mask the interpretation of the results
- Unable to return for follow-up
- Underlying diseases including asthma, heart failure, myocardial infarction, liver failure, renal failure
- Pregnant and lactating women

Ages Eligible for Study
18 Years to 55 Years (Adult)

Sexes Eligible for Study
All

Accepts Healthy Volunteers
No

Design Details
Primary Purpose: Treatment
Allocation: N/A
Interventional Model: Single Group Assignment
Masking: None (Open Label)

Arms and interventions

Participant group/arm	Intervention/treatment
Experimental: stem cells Intravitreal injection of bone marrow-derived CD34+, CD133+, CD271+ stem cells	Biological: stem cell transplantation • Bone marrow-derived CD34+, CD133+, CD271+ stem cells in 1.0 mL normal saline will be injected into the vitreous cavity

Primary outcome measures

Outcome measure	Measure description	Time frame
ETDRS visual acuity change		12 months from baseline

Secondary outcome measures

Outcome measure	Measure description	Time frame
Quality of Life: Questionnaire VFQ-25 (Visual Function Questionnaire-25)		12 months from baseline
Color Vision: Ishihara Color Test		12 months from baseline
Contrast sensitivity: CSV-1000E		12 months from baseline

Sponsor
Stem Cells Arabia

Collaborators
No information provided

Investigators
No information provided

General Publications
No publications available

Korea

Unknown Status

The Safety and Tolerability of Subretinal Transplantation of SCNT-hES-RPE Cells in Patients with Advanced Dry AMD

ClinicalTrials.gov ID NCT03305029

Sponsor CHA University
Information provided by Won Kyung, Song, CHA University (Responsible Party)
Last Update Posted 2017-10-09

Brief Summary
To evaluate the safety and tolerability of human somatic cell nuclear transfer embryonic stem cell-derived retinal pigmented epithelial (SCNT-hES-RPE) cellular therapy in patients with advanced dry AMD.

Detailed Description
To evaluate the safety and tolerability of the SCNT-hES-RPEs, to assess the number of SCNT-hES-RPE cells to be transplanted in future studies, and to evaluate on an exploratory basis potential efficacy endpoints to be used in future studies of SCNT-hES-RPE cellular therapy.

Official Title
Open-Label, Single-Center, Prospective Study to Determine the Safety and Tolerability of Subretinal Transplantation of SCNT-hES-RPE Cells in Patients With Advanced Dry AMD

Conditions
Dry Age-Related Macular Degeneration

Intervention/Treatment
Drug: SCNT-hES-RPE Cells

Study Start
2016-05

Primary Completion (Estimated)
2019-04

Study Completion (Estimated)
2019-04

Enrollment (Estimated)
3

Study Type
Interventional

Phase
Phase 1

Other Study ID Numbers
CHA2015-08-141

Korea, Republic of
Gyeonggi-do Locations

Seongnam-si, Gyeonggi-do, Korea, Republic of, 463-712
CHA Bundang Medical Center

Inclusion Criteria
- Adult male or female over 50 years of age.
- Patient should be in sufficiently good health to reasonably expect survival for at least 4 years after treatment.
- Clinical findings consistent with advanced dry AMD with evidence of one or more areas of >250 μm of geographic atrophy (as defined in the Age-Related eye Disease Study [AREDS] study) involving the central fovea.
- GA defined as attenuation or loss of RPE as observed by biomicroscopy, OCT, and FA.
- The visual acuity (BCVA) of the eye to receive the transplant will be no better than 20/200 in Early Treatment Diabetic Retinopathy Study (ETDRS) visual acuity.
- The visual acuity (BCVA) of the eye that is NOT to receive the transplant will be same or better than the eye to receive the transplant.
- Medically suitable to undergo vitrectomy and subretinal injection.
- Medically suitable for general anesthesia or waking sedation, if needed.
- If female and of childbearing potential, willing to medically acceptable methods of contraception during the study.
- If male, willing to use barrier and spermicidal contraception during the study.
- Willing to defer all future blood, blood component, or tissue donation.
- Able to understand and willing to sign the informed consent.

Exclusion Criteria
- Presence of active or inactive CNV in the eye to be treated.
- Presence or history of retinal dystrophy, retinitis pigmentosa, chorioretinitis, central serious choroidopathy, diabetic retinopathy, other retinal vascular or degenerative disease other than ARMD, optic neuropathy, uveitis, intraocular inflammatory disease, retinal detachment repair, or any other sight-threatening ocular disease.

- Presence of glaucomatous optic neuropathy in the study eye, uncontrolled IOP, or use of two or more agents to control IOP (acetazolamide, beta blocker, alpha-1-agonist, prostaglandins, anhydrous carbonic inhibitors).
- Cataract of sufficient severity likely to necessitate surgical extraction within 1 year.
- Axial myopia of greater than −8 diopters.
- Axial length greater than 28 mm.
- Presence of significant lens opacities or other media opacity.
- History of ocular lens removal within previous 3 months in the study eye.
- History of ocular surgery in the study eye in the previous 3 months in the study eye.
- History of malignancy or evidence of malignancy in screening test.
- Medically not suitable for transplantation of an embryonic stem cell line: Any laboratory value which falls slightly outside of the normal range will be reviewed by the Medical Monitor and Investigators to determine its clinical significance. If it is determined not to be clinically significant, the patient may be enrolled into the study.
  - History of drug abuse, identified in medical history taking.
  - Serologic evidence of infection with Hepatitis B, Hepatitis C, or HIV.
  - Any immunodeficiency.
  - Negative cancer screening within previous 12 months: complete history and physical examination; negative chest roentgenogram (CXR); negative blood test (including CBC and manual differential); negative urinalysis (U/A); normal thyroid exam (T3, T4, TSH); if male, negative for prostate specific antigen (PSA); negative for upper gastrointestinal series or esophagogastroduodenoscopy; negative for α-fetoprotein (AFP); negative fecal occult blood test and negative colonoscopy; if female, normal clinical breast exam and, negative mammogram, negative breast ultrasonography; if female, normal pelvic examination with Papanicolaou smear.
  - Any of the below criteria of diabetes: Fasting blood glucose $\geq$126 mg/dL; random plasma glucose with typical symptoms of diabetes $\geq$200 mg/dL; HbA1c $\geq$7%.
- History of myocardial infarction or cerebrovascular accident in previous 12 months.
- History of cognitive impairments or dementia which may impact the patient's ability participate in the informed consent process and to appropriately complete evaluations.
- Current participation in any other clinical trial.
- Participation within previous 6 months in any clinical trial of a drug by ocular or systemic administration.
- If female, pregnancy or lactation.
- Any other medical condition, which, in the Investigator's judgment, will interfere with the patient's ability to comply with the protocol, compromises patient safety, or interferes with the interpretation of the study results.

Ages Eligible for Study
50 Years and older (Adult, Older Adult)

Sexes Eligible for Study
All

Accepts Healthy Volunteers
No
How is the study designed?

Primary Purpose: Treatment
Allocation: N/A
Interventional Model: Single Group Assignment
Masking: None (Open Label)

Arms and interventions

Participant group/arm	Intervention/treatment
Experimental: SCNT-hES-RPE cells Pars plana vitrectomy and subretinal transplantation of human somatic cell nuclear transfer embryonic stem cell-derived retinal pigmented epithelial cells (SCNT-hES-RPE cells) in patients with advanced dry age-related macular degeneration (AMD)	Drug: SCNT-hES-RPE cells • Pars plana vitrectomy and subretinal transplantation of human somatic cell nuclear transfer embryonic stem cell-derived retinal pigmented epithelial (SCNT-hES-RPE) cells in patients with advanced dry age-related macular degeneration

Primary outcome measures

Outcome measure	Measure description	Time frame
Safety of SCNT-hES-RPE cells	The transplantation of SCNT-hES-RPE cells will be considered safe in the absence of: 1. Any grade 2 (NCI CTCAE V4.03) or greater adverse event related to the cell product 2. Any evidence that the cells are contaminated with an infectious agent 3. Any evidence that the cells show tumorigenic potential	60 months

Secondary outcome measures

Outcome measure	Measure description	Time frame
• Change in the mean of BCVA	Best-corrected ETDRS visual acuity scores	60 months
• Autofluorescence photography	Autofluorescence photography of GA	60 months
• Reading speed	MN reading speed measurements	60 months
• Structural evidence (OCT imaging, fluorescein angiography, autofluorescence photography, slit-lamp examination with fundus photography) that cells have been implanted in the correct location	Evidence of structural changes	60 months

Sponsor
CHA University

Collaborators
No information provided

Investigators
No information provided

General Publications
No publications available

Lebanon

Unknown Status

Autologous Adipose-Derived Adult Stem Cell Transplantation for Corneal Diseases (A-ADAS-CT-CD)

ClinicalTrials.gov ID NCT02932852

Sponsor Vissum, Instituto Oftalmológico de Alicante
Information provided by Vissum, Instituto Oftalmológico de Alicante (Responsible Party)
Last Update Posted 2016-10-13

Brief Summary
Transplantation of cellularized human cornea impregnated and populated by mesenchymal stem cells derived from the patient's adipose tissue. The purpose of the study is to assess the safety, tolerability, and preliminary efficacy of transplantation of a single dose of autologous mesenchymal adipose tissue-derived adult stem cells (ADAS) cellularized into laminas for subjects with corneal defects. Three groups will be included in the study: (1) transplantation of ADAS alone without scaffold, (2) transplantation of scaffold (human corneal decellularized lamina) without ADAS, and (3) transplantation of ADAS cellularized on scaffold (the human corneal decellularized lamina).

Detailed Description
Different types of stem cells have been used in various ways in several research projects in order to find the optimal procedure to regenerate the human corneal stroma. It included several approaches which can be classified as intrastromal implantation of stem cells (1) alone, (2) together with a biodegradable scaffold, (3) with a nonbiodegradable scaffold, or (4) with a decellularized corneal stromal scaffold. In fact, the complex structure of the corneal stroma has not been yet replicated, and there are well-known drawbacks to the use of synthetic scaffold-based designs.

Recently, several corneal decellularization techniques have been described, which provide an acellular corneal extracellular matrix (ECM). These scaffolds have gained attention in the last few years as they provide a more natural environment for the growth and differentiation of cells when compared with synthetic scaffolds. In addition, components of the ECM are generally conserved among species and are tolerated well even by xenogeneic recipients.

Keratocytes are essential for remodeling the corneal stroma and for normal epithelial physiology. This highlights the importance of transplanting a cellular substitute together with the structural support (acellular ECM) to undertake these critical functions in corneal homeostasis. To the best of the investigator's knowledge, all attempts to repopulate decellularized corneal scaffolds have used corneal cells, but these cells have major drawbacks that preclude their autologous use in clinical practice (damage of the donor tissue, lack of cells and inefficient cell subcultures), thus the efforts to find an extraocular source of autologous cells. A recent study by the investigators has shown a perfect biointegration of human decellularized corneal stromal sheets (100 μm thickness) with and without h-ADASC colonization inside the rabbit cornea in vivo, without observing any rejection response despite the graft being xenogeneic. The investigators also demonstrated the differentiation of h-ADASCs into functional keratocytes inside these implants in vivo, which then achieved their proper biofunctionalization. According to the investigator's opinion the transplant of stem cells together with decellularized corneal ECM would be the best technique to effectively restore the thickness of a diseased human cornea, like in keratoconus.

Through this technique, and using extraocular mesenchymal stem cells from patients, it is possible to transform allergenic grafts into functional autologous grafts, theoretically avoiding the risk of rejection.

The process flow is defined as following: (1) the file study: which start by receiving the file of the patient, the file will be forwarded to appointed physicians coordinators for review and submission of medical report, then the medical report will be evaluated within the cell therapy committee and the patient will be asked for clinical examination, and after consultation a reply to the patient with medical decision will follow with an approval or not to be recruited and if yes, a brief report about the procedure will be submitted and explained in details to the patient, a consent form must be signed if the patient agree to be included in the study, (2) the patient admission: which may start by completion of the procedure forms and doing the pre-op evaluation (initial workup defined as a clinical and biological assessment upon C.A.S which may include unaided and best spectacle corrected visual acuity, refraction, slit-lamp examination, intraocular pressure, fundoscopy, corneal topography, aberrometry, endothelial cell count (specular microscopy) as well as blood tests) by an ophthalmologist, followed by lipoaspiration of the subcutaneous adipose tissue to be performed by a plastic surgeon, then sample processing at a cGMP facility for isolation and characterization of the stromal vascular fraction enriched with ADAS as well as laminas preparing (the quality control assessment will be realized before and after all the steps of the procedure starting from the ADAS collection to transplantation including cell culture, cell quiescence, decellularization of human corneal lamina, recellularization of the laminas with ADAS, and transplantation; the assessment will include cell viability, cell number, cell apoptosis, immunophenotyping, infection, inflammation by analyzing the secreted cytokines from ADAS, lamina cutting, lamina thickness, immunostaining, confocal microscopy, etc.); (3) the delivery (transplantation) which starts by a peribulbar or retrobulbar anesthesia where the patient is placed under operating microscope, followed by a femtosecond laser-assisted mid-stromal lamellar dissection, then the ADAS or laminas or recellularized laminas will be placed within the pocket, and finally closed by a superior incision closure. The patient will be put under antibiotic/steroids for a defined period and followed by the team at 1 week, 2 weeks, 1 month, 3 months, 6 months, and 1 year.

This protocol of cell therapy will be applied exclusively at Laser Vision (Hazmieh-Lebanon) and prepared at Reviva Regenerative Medicine Center in affiliation with VISSUM (Alicante-Spain) and Optica General (Saida-Lebanon).

The patient's recruitment will take place: for the lipoaspiration, processing, preparing the cellularized laminas, and quality control assessments, at the Middle East Institute of Health the Reviva Center (Bsalim-Lebanon) under the regulations of the hospital and the Reviva Center; for injection, at Laser vision (Hazmieh-Lebanon); and for follow-up, at Laser vision and optica general. All these steps are managed by VISSUM (Prof. J. ALIO).

Official Title
Autologous Adipose-Derived Adult Stem Cell Transplantation for Corneal Diseases

Conditions
Hereditary Corneal Dystrophy
Keratoconus

Study Type
Interventional

Enrollment (Estimated)
12

Intervention/Treatment
Procedure: Lipoaspiration
Procedure: Transplantation

Phase
Early Phase 1

Other Study ID Numbers
CTP-007/15

Study Start
2015-11

Primary Completion (Estimated)
2017-04

Study Contact
Name: Jorge Alio Sanz, MD, PhD
Phone Number: +34902333344
Email: jlalio@vissum.com

Study Contact Backup
Name: Nehman Makdissy, PhD
Phone Number: +96171210250
Email: almakdissy@hotmail.com
Lebanon

Saida, Lebanon, 10450
Recruiting
Optica General

Contact
Mona El Zarif, OD
+9613610200 monazarifaj@hotmail.com

Contact
Ziad Abduljawad, OD
+9613337335 zicco66@hotmail.com

Eligibility Criteria
Ages Eligible for Study
18 Years and older (Adult, Older Adult)

Accepts Healthy Volunteer
No

Sexes Eligible for Study
All

Inclusion Criteria
- Patients affected by corneal stroma dystrophies of any type, but particularly keratoconus, showing clear evidence at the ophthalmic examination of the presence and clear expression of the disease
- Age: ≥18 years
- Gender: any
- Willing to collaborate and to attend to the clinical follow-ups for the next 5 years
- Patients willing to sign informed consent
- Able and willing to comply with all study requirements
- Patients with and no worse than 0.6 for the better vision patients
- Comprehensive clinical ophthalmological including slit-lamp photography of the cornea of either eye
- Corneal topography map including both anterior and posterior corneal surfaces
- Specular microscopy
- Confocal microscopy of the center of the cornea
- Absence of anterior segment inflammation

- Medically suitable to undergo corneal graft surgery with local anesthesia
- Medically suitable for general anesthesia or waking sedation, if needed
- Normal serum chemistry and hematology screening tests
- Negative human immunodeficiency virus (HIV), hepatitis B (HBV), hepatitis C (HCV) serology
- No history of malignancy
- Complete history and physical examination
- Negative chest roentgenogram (CXR)
- Negative urinalysis (U/A)
- Normal thyroid exam

Exclusion Criteria

- Patients not willing to sign informed consent
- Defects in corneal transparency with a potential to affect the visual outcome
- Vision of 0.05 or less
- Severe corneal deformation
- Infection
- Moderate or severe dry eye
- Chronic ocular surface inflammation of any origin
- Previous ocular surgery other than cataract
- Presence of cataract or other media opacity that would influence ocular fundus documentation and adequate ERG and visual field evaluation
- Other ophthalmic comorbidity like glaucoma and uveitis
- Known allergy: example to proparacaine, ciprofloxacin, fluorescein, indocyanine green, or others to specify
- Known coagulation abnormalities
- Any medical condition likely interferes to cause serious adverse events during the study
- Presence of active or inactive choroidal neovascularization (CNV) in the eye to be treated
- History of malignancy
- History of cognitive impairments or dementia which may impact the patient's ability to participate in the informed consent process and to appropriately complete evaluations
- Any immunodeficiency
- Any current immunosuppressive therapy other than intermittent or low-dose corticosteroids
- Renal insufficiency, as defined by creatine level >1.3 mg/dL
- Serologic evidence of infection with Hepatitis B, Hepatitis C, or HIV
- If female, pregnancy or lactation
- Any other medical condition, which, in the Investigator's judgment, will interfere with the patient's ability to comply with the protocol, compromises patient safety, or interferes with the interpretation of the study results, corneal dystrophies of any type, in a stage which either could be prevented in progression (with best-corrected visual acuity superior to 0.6) or with indication for keratoplasty

(best-corrected visual acuity <0.2)The cases should not have defects in corneal transparency which would lead to vision of 0.05 or less or severe corneal optical deformation (in the case of keratoconus)
- The visual acuity of the eye to receive the transplant will be no better than 20/400
- The visual acuity of the eye in the better vision cohort to receive the transplant will be no better than 0.1 at the Snellen charts
- The visual acuity of the eye that is not to receive the transplant will be no better than 0.2 for the worse vision patients

Design Details
Primary Purpose: Treatment
Allocation: Non-Randomized
Interventional Model: Parallel Assignment
Masking: None (Open Label)

Arms and interventions

Participant group/arm	Intervention/treatment
Active comparator: Group I Lipoaspiration and transplantation of ADAS alone without scaffold	Procedure: lipoaspiration • The patient will have a liposuction surgery from which the autologous ADAS cell to transplant in the cornea will be obtained Procedure: transplantation • The ophthalmologist will carry out the transplantation of the ADAS cells with or without scaffold in the intrastromal cornea of the patient
Active comparator: Group II Lipoaspiration and transplantation of scaffold (human corneal decellularized lamina) without ADAS	Procedure: lipoaspiration • The patient will have a liposuction surgery from which the autologous ADAS cell to transplant in the cornea will be obtained Procedure: transplantation • The ophthalmologist will carry out the transplantation of the ADAS cells with or without scaffold in the intrastromal cornea of the patient
Active comparator: Group III Lipoaspiration and transplantation of ADAS cellularized on scaffold (the human corneal decellularized lamina)	Procedure: lipoaspiration • The patient will have a liposuction surgery from which the autologous ADAS cell to transplant in the cornea will be obtained Procedure: transplantation • The ophthalmologist will carry out the transplantation of the ADAS cells with or without scaffold in the intrastromal cornea of the patient

Primary outcome measures

Outcome measure	Measure description	Time frame
Vision recovery (corrected distance visual acuity)	The best-corrected visual acuity will be measured in each postoperative visit to control any important decrease relative to the surgery	Postoperative visit at 1 day, 1 month, 3 months, and 6 months

Secondary outcome measures

Outcome measure	Measure description	Time frame
Corneal thickness	At each postoperative visit, a measurement of corneal pachymetry of the transplanted cornea will be carried out to detect any abnormal evolution	Postoperative visit at 1 day, 1 month, 3 months, and 6 months
Topography	At each postoperative visit, a measurement of anterior surface topography of the transplanted cornea will be carried out to detect any abnormal evolution	Postoperative visit at 1 month, 3 months, and 6 months
Anterior segment optical coherence tomography	At each postoperative visit, a measurement of the anterior segment by optical coherence tomography of the transplanted cornea will be carried out to detect any abnormal evolution	Postoperative visit at 1 month, 3 months, and 6 months
Slit-lamp observation	At each postoperative visit, the corneal aspect relative to new tissue will be evaluated by slit-lamp observation	Postoperative visit at 1 day, 1 month, 3 months, and 6 months
Refraction measurement	At each postoperative visit, the corneal aspect relative to increase in irregular astigmatism will be evaluated by refraction measurement	Postoperative visit at 1 day, 1 month, 3 months, and 6 months

Sponsor
Vissum, Instituto Oftalmológico de Alicante

Collaborators
- Reviva Pharmaceuticals
- Hospital Universitario La Paz
- Laser Vision Center in Lebanon
- Optica General in Lebanon

Investigators
- Study Chair: Mona El Zarif, OD, Optica General

General Publications

- Alio del Barrio JL, Chiesa M, Garagorri N, Garcia-Urquia N, Fernandez-Delgado J, Bataille L, Rodriguez A, Arnalich-Montiel F, Zarnowski T, Alvarez de Toledo JP, Alio JL, De Miguel MP. Acellular human corneal matrix sheets seeded with human adipose-derived mesenchymal stem cells integrate functionally in an experimental animal model. Exp Eye Res. 2015 Mar;132:91–100. https://doi.org/10.1016/j.exer.2015.01.020. Epub 2015 Jan 24.
- De Miguel MP, Alio JL, Arnalich-Montiel F, Fuentes-Julian S, de Benito-Llopis L, Amparo F, Bataille L. Cornea and ocular surface treatment. Curr Stem Cell Res Ther. 2010 Jun;5(2):195–204. https://doi.org/10.2174/157488810791268663.
- Alio del Barrio JL, Chiesa M, Gallego Ferrer G, Garagorri N, Briz N, Fernandez-Delgado J, Sancho-Tello Valls M, Botella CC, Garcia-Tunon I, Bataille L, Rodriguez A, Arnalich-Montiel F, Gomez Ribelles JL, Antolinos-Turpin CM, Gomez-Tejedor JA, Alio JL, De Miguel MP. Biointegration of corneal macroporous membranes based on poly(ethyl acrylate) copolymers in an experimental animal model. J Biomed Mater Res A. 2015 Mar;103(3):1106–18. https://doi.org/10.1002/jbm.a.35249. Epub 2014 Jun 18.

Malaysia

Unknown Status

Transplantation of Cultivated Corneal Epithelial Sheet in Patients with Ocular Surface Disease (CLET)

ClinicalTrials.gov ID NCT01123044

Sponsor Ministry of Health, Malaysia
Information provided by Ministry of Health, Malaysia
Last Update Posted 2011-07-08

Brief Summary
The aim of the study is to assess if transplantation of cultivated corneal epithelial stem cells could restore vision in patients with severe ocular surface disorder with a favorable safety profile that warrants further comparative study.

Detailed Description
To determine the efficacy of cultivated corneal epithelial stem cells as a treatment for patients with severe ocular surface disorder.

The corneal limbal epithelial stem cell transplant (CLET) successful outcome will be measured by improvement of vision, maintenance of corneal re-epithelialsation with absence of recurrence of surface disease and subjective improvement of symptoms.

Official Title
Transplantation of Cultivated Corneal Epithelial Sheet in Patients With Ocular Surface Disease

Conditions
Eye Injury

Study Type
Interventional

Enrollment (Estimated)
42

Intervention/Treatment
Biological: conservative
Procedure: Medical Therapy

Phase
Phase 3

Other Study ID Numbers
CT 09-01

Study Start
2011-08

Primary Completion (Estimated)
2012-06

Study Completion (Estimated)
2012-09

Enrollment (Estimated)
42

Study Type
Interventional

Phase
Phase 3

Other Study ID Numbers
CT 09-01

Malaysia
Wilayah Persekutuan Locations

Kuala Lumpur, Wilayah Persekutuan, Malaysia, 50586
Hospital Kuala Lumpur

Contact
Umapathy
thiagesw@yahoo.com

Principal Investigator
Thiages Umapathy, Dr.

Ages Eligible for Study
18 Years to 75 Years (Adult, Older Adult)

Sexes Eligible for Study
All

Accepts Healthy Volunteers
No

Eligibility Criteria
Inclusion Criteria

- Male or nonpregnant females between 18 and 75 years of age.
- Written informed consent obtained from patient or parents/guardian.
- Patients with unilateral limbal stem deficiency 2–12 months presenting with any of the following
  - Conjunctivalization
  - Absence of limbal palisades of Vogts
  - Chronic inflammation
  - Persistent or recurrent corneal epithelial defect

- Patients who had 2–12 months of conservative treatment. Diagnostic criteria for limbal stem cell deficiency are as follows:
  - Symptoms of decreased vision, redness, watering, photophobia, and recurrent attacks of pain
  - Triad of signs: conjunctivalization, neovascularization, and chronic inflammation
  - Stippled appearance of conjunctivalized cornea with loss of palisades of Vogt
  - Recurrent and persistent epithelial defects
  - Superficial vascularization, scarring, thick fibrovascular pannus, ulceration, melting, and perforation

Exclusion Criteria
Patients with any of the following are not eligible for enrollment into the study:

- Pregnant or nursing woman or women of childbearing potential except if postmenopausal, surgically sterile or using accepted method(s) of birth control, or having negative pregnancy test.
- Participation in any drug trial in which the patient received an investigational drug within 30 days or 5 half-lives (whichever is longer) preceding the screening phase of this study.
- Those persons directly involved in the conduct of the study.
- Any history of severe infection such as hepatitis, renal, gastrointestinal, endocrine, or neurological disease.
- Evidence of corneal stromal scarring, cataract, macular oedema or scarring, retinal detachment, and conjunctival keratinization.
- Positive for HIV, Hepatitis B, C, and VDRL.
- History of Pulmonary tuberculosis, hepatitis B.
- History of alcohol or substance abuse.
- History of malignancy within previous 5 years.
- History of organ transplant.
- Any serious medical conditions or disability, which in the opinion of the investigator, would interfere with treatment or assessment or preclude completion of this study.

Ages Eligible for Study
18 Years to 75 Years (Adult, Older Adult)

Sexes Eligible for Study
All

Accepts Healthy Volunteers
No

Design Details
Primary Purpose: Treatment
Allocation: Randomized
Interventional Model: Parallel Assignment
Masking: None (Open Label)

Arms and interventions

Participant group/arm	Intervention/treatment
Experimental: corneal stem cell transplant	Biological: conservative • Undergo autologous transplantation of limbal epithelial cells cultured on amniotic membrane
No intervention: conservative medical therapy	Biological: conservative • Undergo autologous transplantation of limbal epithelial cells cultured on amniotic membrane Procedure: medical therapy • Under usual care treatment

Primary outcome measures

Outcome measure	Measure description	Time frame
Improvement of vision at 1 week	The primary efficacy variable will be improved visual acuity vision will be assessed using ETDRS logMAR visual acuity measurements. An improvement of >1 line	1 week
Improvement of vision at 1 month	The primary efficacy variable will be improved visual acuity vision will be assessed using ETDRS logMAR visual acuity measurements. An improvement of >1 line	1 month
Improvement of vision at 3 month	The primary efficacy variable will be improved visual acuity vision will be assessed using ETDRS logMAR visual acuity measurements. An improvement of >1 line	3 month
Improvement of vision at 6 month	The primary efficacy variable will be improved visual acuity vision will be assessed using ETDRS logMAR visual acuity measurements. An improvement of >1 line	6 month
Improvement of vision at 9 month	The primary efficacy variable will be improved visual acuity vision will be assessed using ETDRS logMAR visual acuity measurements. An improvement of >1 line	9 month
Improvement of vision at 12 month	The primary efficacy variable will be improved visual acuity vision will be assessed using ETDRS logMAR visual acuity measurements. An improvement of >1 line	12 month

Secondary outcome measures

Outcome measure	Measure description	Time frame
Adverse events reporting, vital signs, and physical examinations	To evaluate the safety and tolerability of transplantation of limbal epithelial cells cultured on amniotic membrane for a follow-up period of 1 year in patients. Safety will be evaluated using adverse events reporting, vital signs, and physical examinations	
Maintenance of corneal re-epithelialsation with absence of recurrence of surface disease		
Subjective improvement of symptoms		

Sponsor
Ministry of Health, Malaysia

Collaborators
No information provided

Investigators
No information provided

General Publications
No publications available

Poland

Unknown Status

Stem Cells Therapy in Degenerative Diseases of the Retina

ClinicalTrials.gov ID NCT03772938

Sponsor Pomeranian Medical University Szczecin
Information provided by Marta P. Wiącek, Pomeranian Medical University Szczecin (Responsible Party)
Last Update Posted 2018-12-17

Brief Summary
The purpose of this study is to test the safety and effectiveness of an autologous bone marrow-derived stem/progenitor cells administered intravitreously in the subjects with degenerative diseases of the retina.

Detailed Description
Degenerative diseases of the retina are challenging for ophthalmologists. This is a common term that covers heterogeneous group of diseases, i.e., retinitis pigmentosa, Stargardt disease, Best's disease, or age-related macular degeneration. Undetermined etiology, wide range of factors that may trigger the onset of the disease and modulate its course impede the implementation of an effective treatment. Currently, a stem cells therapy seems to be promising option in patients with degenerative diseases of the retina. The purpose of this prospective, nonrandomized, open-label, pilot study is to conduct the investigation of the safety and efficacy of intravitreal injection of autologous bone marrow-isolated stem/progenitor cells with different selected phenotypes into the subjects with degenerative diseases of the retina. Especially, this clinical trial is designated to test the therapeutic (pro-regenerative and neuro-protective) functions of different stem/progenitor cell populations able to secrete bioactive neurotrophic factors. All patients enrolled will have a documented history of degenerative disease of the retina prior to study enrollment.

Next, autologous bone marrow-isolated stem/progenitor cells intravitreous injection will be performed.

Finally, treatment safety, adverse events, and exploratory parameters, including best-corrected visual acuity, visual field, and electroretinography parameters, to establish disease progression rate will be recorded throughout the duration of the posttreatment follow-up period.

Official Title
Stem Cells Therapy in Degenerative Diseases of the Retina

Conditions
Retinal Degeneration
Retinitis Pigmentosa
Age-Related Macular Degeneration
Stargardt Disease

Study Type
Interventional

Enrollment (Estimated)
30

Intervention/Treatment
Biological: Stem/progenitor cells transplantation

Phase
Phase 1

Other Study ID Numbers
IKiKO-KB-0012/143/13

Study Start (Actual)
2018-12-13

Primary Completion (Estimated)
2020-02-28

Study Completion (Estimated)
2020-03-31

Poland

Szczecin, Poland, 70-111
I Department of Ophthalmology

Eligibility Criteria
Ages Eligible for Study
18 Years to 65 Years (Adult, Older Adult)

Accepts Healthy Volunteer
No

Sexes Eligible for Study
All

Inclusion Criteria
- Diagnosed degenerative disease of the retina,
- Age 18–65 years,
- Best-corrected visual acuity max. 0.2 (Snellen letter chart),
- Good understanding of the protocol and willingness to consent,
- Signed informed consent.

Exclusion Criteria
- Concomitant eye disease (glaucoma, etc.)
- Concomitant of other systemic disease or diseases,
- Inflammation (high protein or lymphocytosis in the CSF), active infections,
- Diabetes,
- Cardio-vascular disorders,
- Cancer,
- Autoimmune diseases,
- Renal failure,
- Impaired hepatic function,
- Subject unwilling or unable to comply with the requirements of the protocol,
- Patient has been treated previously with any cellular therapy.

Primary Purpose: Other
Allocation: Non-Randomized
Interventional Model: Parallel Assignment
Masking: None (Open Label)

Arms and interventions

Participant group/arm	Intervention/treatment
Active comparator: stem/progenitor cells transplantation Intervention: a single intravitreal injection of autologous bone marrow-derived stem/progenitor cells will be performed	Biological: stem/progenitor cells transplantation • Human autologous bone marrow-derived stem/progenitor cell transplantation administered as an intravitreal injection in patients with degenerative disease of retina
Sham comparator: standard treatment of degenerative disease of retina Symptomatic treatment of degenerative disease of retina without biologic cell-based treatment	Biological: stem/progenitor cells transplantation • Human autologous bone marrow-derived stem/progenitor cell transplantation administered as an intravitreal injection in patients with degenerative disease of retina

Primary outcome measures

Outcome measure	Measure description	Time frame
Number of participants with treatment-related adverse events as assessed by CTCAE v4.0	Confirm the safety of autologous bone marrow stem/progenitor cell intravitreal injection in enrolled patients by repeated follow-up over 1 year with clinical evaluations	12 months

Secondary outcome measures

Outcome measure	Measure description	Time frame
Efficacy of autologous bone marrow stem/progenitor intravitreal injection in enrolled patients	Best-corrected visual acuity, ETDRS chart [number of letters]	12 months
Intraocular pressure	Pascal tonometer [mmHg]	12 months
Optic disk retinal nerve fiber layer	Optical coherence tomography [μm]	12 months
Central macular thickness	Optical coherence tomography [μm]	12 months
Ganglion cell complex thickness	Optical coherence tomography [μm]	12 months
Choroidal thickness	Enhanced depth imaging optical coherence tomography [μm]	12 months
Choroidal volume	Enhanced depth imaging optical coherence tomography [mm^3]	12 months
Computed perimetry (30-2 and 10-2 module)	Mean deviation, pattern standard deviation [B]	12 months
Goldmann perimetry with color filters	[degrees]	12 months
Contrast sensitivity	Pelli-Robson chart [number of letters]	12 months
Function of the photoreceptors (rods and cones), inner retinal cells (bipolar and amacrine cells), and ganglion cells	Electroretinography (ERG) examination: • amplitude of a and b waves [V] • culmination time of a and b waves [s] • culmination time of q1-q3 waves [s]	12 months
Function of the photoreceptors	Multifocal electroretinography (mfERG) examination: • retinal response density [V/degree 2] • culmination time of P1 wave in 6 rings [s]	12 months
Function of ganglion cells	Pattern electroretinography (PERG) examination: • amplitude of P50 and N95 waves [V] • culmination time of P50 wave [s]	12 months

Sponsor
Pomeranian Medical University Szczecin

Collaborators
No information provided

Investigators
- Study Chair: Bogusław Machaliński, MD, PhD, Pomeranian Medical University
- Study Director: Anna Machalińska, MD, PhD, Pomeranian Medical University

General Publications
No publications available

Spain

Unknown Status

Safety Assessment of Intravitreal Mesenchymal Stem Cells for Acute Non-arteritic Anterior Ischemic Optic Neuropathy (NEUROSTEM)

ClinicalTrials.gov ID NCT03173638

Sponsor Instituto Universitario de Oftalmobiología Aplicada (Institute of Applied Ophthalmobiology) - IOBA
Information provided by Instituto Universitario de Oftalmobiología Aplicada (Institute of Applied Ophthalmobiology) - IOBA (Responsible Party)
Last Update Posted 2021-04-06

Brief Summary
In this study the investigators propose to perform a Phase II Clinical Trial which seeks to evaluate the safety of cell therapy as a new treatment for patients who suffer from acute non-arteritic anterior ischemic optic neuropathy (NAION). If it is successful, a later study to assess efficacy will be performed.

This disease has an age-adjusted incidence rate of 10/100,000, and in many cases it results in blindness. Currently, there is no available treatment and second eye involvement occurs in approximately 15–25% of the cases.

All this background originates a particularly dramatic outcome for the patient. Therefore, it seems justified to evaluate new therapies that maintain or improve the visual function in these patients.

The Project includes a clinical trial whose purpose is the assessment of the safety of intravitreal administration of allogenic mesenchymal stem cells (MSCs) in patients with NAION, a product in clinical research phase (PEI No. 15-007) already approved for other human clinical applications (PEI No. 15-007).

It can be considered that the therapy with intravitreal injection of MSC is a treatment option in patients with acute NAION, since through the paracrine properties of these cells (secretion of neurotrophic, immunomodulatory, and anti-apoptotic factors) it may prevent or reduce the progression of axonal degeneration caused by this disease.

Official Title
Phase II Safety Assessment of Intravitreal Injection of Mesenchymal Stem Cells for Acute Non Arteritic Anterior Ischemic Optic Neuropathy (NAION)

Conditions
Non Arteritic Ischemic Optic Neuropathy

Study Type
Interventional

Enrollment (Actual)
5

Intervention/Treatment
Procedure: intravitreal injection of MSV

Phase
Phase 2

Other Study ID Numbers
IOBA01-2016
2016-003029-40 (EudraCT Number)

Study Start (Actual)
2018-03-23

Primary Completion (Estimated)
2022-04

Study Completion (Estimated)
2022-12

Valladolid, Spain, 47011
IOBA - Applied Ophthalmobiology Institute

Eligibility Criteria
Ages Eligible for Study
18 Years to 80 Years (Adult, Older Adult)

Accepts Healthy Volunteer
No

Sexes Eligible for Study
All

Inclusion Criteria

- Patients with acute unilateral NAION (within 2 weeks after the first symptoms) with at least two of the following:
  - Sudden and unpainful monocular vision loss.
  - Visual field defects.
  - Dyschromatopsia.
  - Ocular nerve head edema.
  - Afferent relative pupil defect.

- Patients ≥50 years old, able to freely give informed consent.
- Best-corrected visual acuity (BCVA) ≤ 0.1 in study eye.
- Pseudophakia in study eye.
- Preserved pupil sphincter muscle motility.
- Signed informed consent form before any study procedure.
- Signed data protection consent form before any study procedure.

Exclusion Criteria

- Giant cell arteritis evidence (clinical history, Erythrocyte sedimentation rate (ESR), C-Reactive Protein).
- Evidence of any other etiology that may justify the optic neuropathy (even in the non-study eye).
- History of systemic vasculitis, multiple sclerosis, collagenopathies, or previous cancer treatments.
- Hypersensitivity or allergy to any compound used in the study, including investigational medicinal product (IMP).
- Positive pregnancy test at baseline.
- Participation in any other research study within 2 months.

Ophthalmic Exclusion Criteria

- History of uveitis or active ocular inflammation.
- History or evidence of glaucoma or high intraocular pressure (≥24 mmHg in either eye).
- Mean opacities or retinal pathologies in the study eye.
- Any previous vitreous or glaucoma surgery in the study eye.
- Cataract surgery within 3 months in the study eye.

Design Details

Primary Purpose: Treatment
Allocation: N/A
Interventional Model: Single Group Assignment
Masking: None (Open Label)

Arms and interventions

Participant group/arm	Intervention/treatment
Experimental: mesenchymal stem from valladolid (MSV) Allogenic mesenchymal stem cells from bone marrow	Procedure: intravitreal injection of MSV • A unique intravitreal injection of MSV in acute phase of NAION

Primary outcome measures

Outcome measure	Measure description	Time frame
Absence of inflammatory reaction—Slit-lamp exploration—SUN (Standardization of Uveitis Nomenclature) scale	Following the SUN scale, the parameters that will be taken into account in the slit-lamp exploration will be: cells and flare in anterior chamber, and vitreous flare	Changes compared to baseline in visits V1 (1 day), V2 (1 week), V3 (1 month), V4 (3 months), V5 (6 months) and V6 (12 months) and in any unscheduled visit any patient might need

Secondary outcome measures

Outcome measure	Measure description	Time frame
Adverse events procedure-related (intravitreal injection)	Conjunctival hemorrhages, anterior chamber inflammation, changes in intraocular pressure, infectious endophthalmitis, vitreous inflammation, retinal detachment, choroidal detachment, corneal opacities, lens opacities, neovascularization, macular edema, or any other adverse event that may appear	Throughout the study after treatment in visits V1 (1 day), V2 (1 week), V3 (1 month), V4 (3 months), V5 (6 months), and V6 (12 months) and in any unscheduled visit any patient might need

Sponsor
Instituto Universitario de Oftalmobiología Aplicada (Institute of Applied Ophthalmobiology) - IOBA

Collaborators
- University of Valladolid

Investigators
- Principal Investigator: José C Pastor Jimeno, MD, PhD, IOBA-UVA

General Publications
No publications available

Taiwan

Unknown Status

The Improvement of Limbal Epithelial Culture Technique by Using Collagenase to Isolate Limbal Stem Cells

ClinicalTrials.gov ID NCT02202642

Sponsor National Taiwan University Hospital
Information provided by National Taiwan University Hospital, National Taiwan University Hospital (Responsible Party)
Last Update Posted 2014-07-29

Brief Summary
Using collagenase to isolate limbal stem cells and improve the technique of ex vivo expansion of limbal stem cells for the treatment of patients suffering from unilateral limbal stem cell insufficiency based on the concept of "limbal stem cells need special cell-cell contact and cell-extracellular matrix interaction to support their survival."

Detailed Description

Patients suffering from unilateral limbal stem cell insufficiency and understand the benefits and risk factors of the operation will be enrolled in this clinical trial. Two weeks before transplantation, the investigators first take 1–2 mm^2 of limbal tissues from the healthy eye, digest the tissue with collagenase, and get an epithelial cell cluster. The cell cluster will be cultured on amniotic membrane with SHEM medium which contains Dulbecco's modified Eagle's medium/F12 (1:1), 5% dimethyl sulfoxide, 10 ng/mL human epidermal growth factor, 5 μg/mL insulin, 0.5 μg/mL hydrocortisone, 1 nM cholera toxin, 50 μg/mL gentamicin, 1.25 μg/mL amphotericin B. Seven to 10 days later after successful culture, transplantation will be performed after the cells reach 1.5–2 cm^2 compact confluent epithelial sheets without evidence of contamination. During the operation, lamellar keratectomy will be performed to remove the abnormal corneal neovascularization and scar tissue. The cultivated epithelial sheets will be transplanted to denuded corneal surface with fibrin glue and suture technique. The patients are planned to be admitted for 2 weeks. After discharge, follow-up of the patients will be performed at 1 week, 2 week, 1 month, 2 months, 3 months, 6 months, and 1 year later.

Pictures of external eyes will be taken and the extent of corneal epithelium regeneration, visual acuity, and complications will be recorded during follow-up visits. The investigators plan to collect ten patients.

Official Title

The Improvement of Limbal Epithelial Culture Technique for the Treatment of Unilateral Limbal Insufficiency by Using Collagenase to Isolate Limbal Stem Cells

Conditions

Alkaline Chemical Burn Of Cornea And Conjunctival Sac
Acid Chemical Burn Of Cornea And Conjunctival Sac
Benign Mucous Membrane Pemphigoid With Ocular Involvement

Study Type

Interventional

Enrollment (Estimated)

10

Intervention/Treatment

Procedure: collagenase

Phase

Phase 1

Other Study ID Numbers

201007032D

Study Start

2012-01

Primary Completion (Estimated)
2015-12

Study Completion (Estimated)
2015-12

Study Contact
Name: Wei Li Chen, MD, PhD
Phone Number: 886-2-23123456 ext 5206
Email: chenweili@ntu.edu.tw
Taiwan

Taipei, Taiwan, 100
Recruiting
Wei-Li Chen

Eligibility Criteria
Ages Eligible for Study
18 Years to 70 Years (Adult, Older Adult)

Accepts Healthy Volunteer
No

Sexes Eligible for Study
All

Inclusion Criteria
- Age: between 18 and 70 years old.
- Lesion eye should have limbal insufficiency more than 180°.
- Lesion eye should have recurrent corneal erosions, corneal neovascularization and poor vision caused by limbal insufficiency.
- The symptoms should be more than 1/2 year, and have no possibility of improvement by medical treatment.
- The patients will to receive the operation and the long term postoperative follow-up.
- The patients are not infected by AIDS, hepatitis B/C, and willing to receive the associated examinations.
- The patients does not plan to be pregnant from the day of cell culture and transplantation for 1 year. They also agree to receive the pregnant test.
- The condition of cell culture from the first to the clinical trial is successful.

Exclusion Criteria
- Bilateral limbal insufficiency.
- No recurrent corneal erosion, corneal neovascularization, or poor vision was found due to limbal insufficiency. If corneal neovascularization was found deeper than the anterior 1/2 of corneal stroma, or corneal thickness was less than 200 μm evaluated by anterior segment OCT and ultrasound pachymetry, the patients should be excluded from the operation criteria. If corneal thickness becomes thicker, the patients can be enrolled for operation.

- About corneal sedation, if the value checked by Cocet-Bonnet esthesiometer is less than 5 mm, the patients should be excluded from operation.
- The condition can be improved by medication, or resolve spontaneously.
- Postoperative follow-up is less than 1/2 year.
- The patient can't receive long-term postoperative follow-up.
- No light perception, or can be expected to have very poor prognosis.
- If the intraocular pressure was more than 21 mmHg under necessary glaucoma surgery and anti-glaucoma medication, or optic disc cupping was more than 90%, or visual field was found to have severe defects, the patients were excluded from the clinical trial.
- Severe lagophthalmos or trichiasis, and does not received blepharoplasty.
- Severe dry eye syndrome. Those patients with Schirmer's test result less than 1 mm should be excluded.
- If the cells in the first part of the clinical trial can't be successfully cultivated, the patients should be excluded. If the patients insisted to received further treatment, they can be enrolled 3 months later.
- Pregnancy

Design Details
Primary Purpose: Treatment
Allocation: N/A
Interventional Model: Single Group Assignment
Masking: None (Open Label)

Arms and interventions

Participant group/arm	Intervention/treatment
Experimental: limbal stem cells Using collagenase to isolate limbal stem cells and improve the technique of ex vivo expansion of limbal stem cells for the treatment of patients suffering from unilateral limbal stem cell insufficiency based on the concept of "limbal stem cells need special cell-cell contact and cell-extracellular matrix interaction to support their survival"	Procedure: collagenase • Cultured limbal stem cells transplantation

Primary outcome measures

Outcome measure	Measure description	Time frame
Using collagenase to isolate limbal stem cells and improve the technique of ex vivo expansion of limbal stem cells for the treatment	Nine to 12 days later after successful culture, transplantation will be performed after the cells diameter reaches 1.5–2.0 cm compact confluent epithelial sheets without evidence of contamination	12 months

Sponsor
National Taiwan University Hospital

Collaborators
No information provided

Investigators

- Principal Investigator: Wei Li Chen, MD, PhD, National Taiwan University Hospital

General Publications

No publications available

Thailand

Unknown Status

Ocular Surface Reconstruction with Cultivated Autologous Mucosal Epithelial Transplantation

ClinicalTrials.gov ID NCT01942421

Sponsor Mahidol University
Information provided by Pinnita Prabhasawat, Mahidol University (Responsible Party)
Last Update Posted 2013-09-16

Brief Summary

The purpose of this study is to elucidate the appropriate condition of developing cultivated autologous oral mucosal epithelial graft and evaluate the surgical outcome of transplantation of the cultivated cells in the patients.

Detailed Description

Ocular surface damage caused by severe ocular surface diseases. Corneal limbal stem cell transplantation has been demonstrated to improve the outcome of ocular surface reconstruction. However, the conventional management of allograft limbal transplantation generally has unsatisfactory outcome because of high risk of rejection and requires long-life immunosuppressive medications. Ocular surface reconstruction with cultivated autologous mucosal epithelial transplantation is served to improve the surgical results of ocular surface reconstruction and decrease the use of immunosuppressive medications.

Official Title

Ocular Surface Reconstruction With Cultivated Autologous Mucosal Epithelial Transplantation

Conditions

Ocular Surface Disease

Study Type

Interventional

Enrollment (Estimated)
20

Intervention/Treatment
Procedure: Cultivated mucosal epithelial transplantation

Phase
Phase 2 Phase 3

Other Study ID Numbers
639/2551(EC2)

Study Start
2008-01

Primary Completion (Actual)
2013-06

Study Completion (Estimated)
2013-12

Bangkok, Thailand, 10300
Pinnita Prabhasawat, MD

Eligibility Criteria
Ages Eligible for Study
18 Years and older (Adult, Older Adult)

Accepts Healthy Volunteer
No

Sexes Eligible for Study
All

Inclusion Criteria
- Bilateral limbal stem cell deficiency proved by specific clinical features and impression cytology

Exclusion Criteria
- Age under 18
- Active ocular infection, Severe dry eye, Uncontrolled glaucoma
- Immunocompromised host
- Pregnancy

Design Details
Primary Purpose: Treatment
Allocation: N/A
Interventional Model: Single Group Assignment
Masking: None (Open Label)

Arms and interventions

Participant group/arm	Intervention/treatment
Experimental: cultivated mucosal epithelial transplant Cultivated mucosal epithelial transplantation to limbal deficiency patients	Procedure: cultivated mucosal epithelial transplantation • Cultivated autologous mucosal epithelium, then transplant to limbal stem cell deficiency patients

Primary outcome measures

Outcome measure	Measure description	Time frame
Ocular surface clinical and pathological status (signs of limbal deficiency and immunopathological markers) after the surgery		3 years

Sponsor

Mahidol University

Collaborators

No information provided

Investigators

- Principal Investigator: Pinnita Prabhasawat, MD, Mahidol University

General Publications

No publications available

Chapter 5
Withdrawn Ocular Stem Cell Studies

United States

Study to Assess the Safety and Effects of Cells Injected Intravitreal in Dry Macular Degeneration

ClinicalTrials.gov ID NCT02024269

Sponsor Bioheart, Inc.
Information provided by Bioheart, Inc. (Responsible Party)
Last Update Posted 2015-09-16

Brief Summary
This will be an open-label, non-randomized multicenter study of adipose stem cell (ASC) implantation. ASCs will be derived from the patient's adipose or fat. Liposuction using local anesthesia and syringe collection will be performed to collect the adipose tissue specimen for subsequent processing to isolate the stem cells. The cells will be delivered via needle injection into the eye.

Official Title
An Open-label, Non-Randomized, Multicenter Study to Assess the Safety and Effects of Autologous Adipose-Derived Stromal Cells Injected Intravitreal in Dry Macular Degeneration

Conditions
Dry Macular Degeneration

Study Type
Interventional

Enrollment (Actual)
0

J. N. Weiss, *Ocular Stem Cell Studies*,
https://doi.org/10.1007/978-3-031-58557-9_5

Intervention/Treatment
Biological: Adipose Derived Stem Cells

Phase
Not Applicable

Other Study ID Numbers
ADI-US-AMD-001

Study Start
2013–12

Primary Completion (Estimated)
2016–12

Study Completion (Estimated)
2017–06

United States

Sunrise, Florida, United States
Bioheart

Eligibility Criteria
Ages Eligible for Study
50 to 90 years (Adult, Older Adult)

Accepts Healthy Volunteer
No

Sexes Eligible for Study
All

Description

Inclusion Criteria
- Males and Females between Age 50 and 90 years
- Patients with dry AMD
- Visual acuity in the study eye <20,200 (equal to or worse than 20/200)

Exclusion Criteria
- Females who are pregnant or nursing or females of childbearing potential who are unwilling to maintain contraceptive therapy for the duration of the study.
- Inability to undergo syringe liposuction procedure or have any medical problems that contraindicate the procedure.
- Life expectancy <6 months due to concomitant illnesses.
- Vitrectomized eyes.
- Exposure to any investigational drug or procedure within 1 month prior to study entry or enrolled in a concurrent study that may confound results of this study.
- Active infectious disease. If patients have tested positive for HIV, HTLV, HBV, HCV, CMV (IgM > IgG), and/or syphilis, an expert will be consulted as to patient eligibility based on the patient's infectious status.

- Any illness which might affect a patient's survival over the follow-up period.
- Any illness which, in the Investigator's judgement, will interfere with the patient's ability to comply with the protocol, compromise patient safety, or interfere with the interpretation of the study results.
- Patients on chronic immunosuppressive transplant therapy.
- Patients with severe coagulation disorders.
- Systolic blood pressure (supine) $\leq$90 mmHg.
- Resting heart rate >100 bpm;.
- Active clinical infection being treated by antibiotics within 1 week of enrollment.
- Cerebrovascular accident within 6 months prior to study entry.
- Known drug or alcohol dependence or any other factors which will interfere with the study conduct or interpretation of the results or who in the opinion of the investigator are not suitable to participate.
- Severe COPD or severe asthma that would contraindicate surgery.
- History of cancer (other than nonmelanoma skin cancer or in situ cervical cancer) in the last 2 years.
- Unwilling and/or not able to give written informed consent.

Design Details

Primary Purpose: Treatment
Allocation: N/A
Interventional Model: Single Group Assignment
Masking: None (Open Label)

Arms and interventions

Participant group/arm	Intervention/treatment
Experimental: Adipose Stem Cells	Biological: Adipose Derived Stem Cells

Primary outcome measures

Outcome measure	Measure description	Time frame
Frequency and nature of adverse events		6 months
Visual Field analysis (10–2)	Visual field testing can be performed clinically by keeping the subject's gaze fixed while presenting objects at various places within their visual field	6 months
Visual acuity measurements	Visual acuity is a measure of the spatial resolution of the visual processing system. VA is tested by requiring the person whose vision is being tested to identify so-called optotypes—stylized letters, Landolt rings, or other patterns—on a chart from a set viewing distance	6 months

Secondary outcome measures

Outcome measure	Measure description	Time frame
Intraocular pressure		6 months

Sponsor

Bioheart, Inc.

Collaborators
No information provided

Investigators
No information provided

General Publications
No publications available

China

Withdrawn

Reason: It was hard to recruit patients.

Treatment of Relapsed and/or Refractory AQP4-IgG Seropositive NMOSD by Tandem CAR T Cells Targeting CD19 and CD20

ClinicalTrials.gov ID NCT03605238

Sponsor Chinese PLA General Hospital
Information provided by Wei Shihui, Chinese PLA General Hospital (Responsible Party)
Last Update Posted 2019-09-19

Brief Summary
CAR-T therapy was proposed and has been recently used for cancer treatment. It has been hailed for its promising remission rates after early-stage clinical trials for acute lymphoblastic leukemia. However, CAR-T therapy is seldom used for autoimmune diseases. Researchers only use it for the treatment of systemic lupus erythematosus (SLE). Neuromyelitis optica spectrum disorders (NMOSD), that include the neuromyelitis optica (NMO), are a group of inflammatory disorders of the central nervous system characterized by episodes of immune-mediated demyelination and axonal damage mainly involving optic nerves and spinal cord. NMO is characterized by the presence of an anti-Aquaporin-4 (AQP4) antibody, which can only be produced by differentiation of B cells to plasma cells. Because these anti-AQP4 antibodies may be pathogenic, B cells recognizing AQP4 may be directly involved in the disease process as well. B cells also play a role as potent antigen presenting cells in NMO. NMO has the characteristics of high recurrence rate and poor prognosis. In the conventional treatment options, NMOSD could be treated with corticosteroids and immunosuppressive drugs immunosuppressant (e.g. azathioprine, mycophenolate mofetil, rituximab). But these drugs could barely completely cure NMOSD. And now, chimeric antigen receptor modified T cell infusion maybe an

effective treatment to solve these problems. The rationale for using CAR-T therapy in NMOSD is based on the known roles of B cells, antibody production, and plasma cells in the pathophysiology of NMOSD. The strongest evidence of the importance of B cells in NMO comes from studies of B cell depletion, most commonly with anti-CD20 monoclonal antibody, rituximab. Emerging evidence indicates that peripheral B cells are activated during a relapse and plasmablast production of anti-AQP4 antibodies spikes. The investigators infuse tanCART19/20 to completely deplete B cells. The purpose of this study is to assess the safety and efficacy of this tanCART19/20 in the treatment of NMOSD.

Detailed Description
This study is being conducted to assess anti-CD19/20 CAR T cells safety and efficacy in treating patients with AQP4-IgG seropositive NMOSD.

Primary Objectives
I. To assess the safety of the tanCART-19/20 cells in treating NMOSD patients. II. Determine duration of in vivo survival of tanCART-19/20 cells.

Secondary Objectives

I. To assess the efficacy of the tanCART-19/20 cells in treating NMOSD patients.
II. The secondary outcome measures: annual relapse rate (ARR), Expanded Disability Status Scale Score (EDDS), Best Corrected Visual Acuity (Log MAR), Spectral-Domain Optical Coherence Tomography (SD-OCT), Flash Visual Evoked Potential (FVEP), and Immunological assessments.

Outline: Patients receive anti-CD19/20-CAR (coupled with CD137 and CD3 zeta signaling domains) vector-transduced autologous T cells on days 0 in the absence of unacceptable toxicity. The infusion dose is 1E5-2E6 CAR positive T cells/kg, and dose escalation methods obey the traditional 3 + 3 design (three doses groups: 1-2E5, 3-6E5, 1-2E6 CAR-T cells).

After completion of study treatment, patients are followed intensively for 6 months, every 6 months for 2 years, and annually thereafter for 3 years.

Official Title
Clinical Study of CD19/CD20 tanCAR T Cells in Relapsed and/or Refractory AQP4-IgG Seropositive Neuromyelitis Optica Spectrum Disorders (NMOSD)
Neuromyelitis Optica Spectrum Disorder

Study Type
Interventional

Enrollment (Actual)
0

Intervention/Treatment
Biological: Corticosteroids & tanCART19/20

Phase
Phase 1

Other Study ID Numbers
CHN-PLAGH-NO-S2018–002-02

Study Start (Estimated)
2018-08-15

Primary Completion (Estimated)
2019-08-15

Study Completion (Estimated)
2020-08-15

China
Beijing, Beijing, China, 100,853
People's Liberation of Army General Hospital (PLAGH)

Eligibility Criteria
Ages Eligible for Study
12 to 75 years (Child, Adult, Older Adult)

Accepts Healthy Volunteer
No

Sexes Eligible for Study
All

Description

Inclusion Criteria
- Clinical diagnosis of neuromyelitis optica spectrum disorders (NMOSD) patients.
- Patients with AQP4-IgG seropositive by cell-based assay.
- Patients with corticosteroid treatment combined immunosuppressant (azathioprine or mycophenolate mofetil or rituximab) still recurrence.
- Clinical evidence of at least 2 relapses in last 12 months or 3 relapses in the last 24 months with at least 1 relapse in the 12 months prior to the Screening.
- Best corrected visual acuity (BCVA) < 20/60.
- Normal bone marrow reserve function: neutrophils>1500/mm^3, Hemoglobin >10 g/dL, Platelet count >100,000/mm^3.
- Normal liver and kidney function: Creatinine <2.5 mg/dL, ALT (alanine aminotransferase)/AST (aspartate aminotransferase) < 3× normal, Bilirubin <2.0 mg/dL.
- Successful test expansion of tanCART19/20 cells.
- Adequate venous access for apheresis, and no other contraindications for leukapheresis.
- Voluntary informed consent is given.

Exclusion Criteria

- Pregnant or lactating women (The safety of this therapy on unborn children is not known, Female study participants of reproductive potential must have a negative serum or urine pregnancy test performed within 48 h before infusion).
- Any serious, uncontrolled diseases (including, but not limit to, uncontrolled active infection, active hepatitis B or hepatitis C infection, HIV infection, unstable angina pectoris, congestive heart failure, serious arrhythmia).
- Concurrent use of systemic steroids or immunosuppressant in the last 2 weeks.
- Feasibility assessment during screening demonstrates <30% transduction of target lymphocytes, or insufficient expansion (<five-fold) in response to CD3/CD137 costimulation.
- Other patients who are not suitable for CAR-T therapy judged by the biotherapy physician.

Design Details

Primary Purpose: Treatment
Allocation: N/A
Interventional Model: Single Group Assignment
Masking: None (Open Label)

Arms and interventions

Participant group/arm	Intervention/treatment
Experimental: Corticosteroids & tanCART19/20 Twelve days of high-dose IV methylprednisolone to reduce acute inflammation, then infuse anti-CD19/20-CAR retroviral vector-transduced autologous derived T cells only once	Biological: Corticosteroids & tanCART19/20 • Twelve days of high-dose IV methylprednisolone (1000 mg × 3 days, 500 mg × 3 days, 240 mg × 3 days, 120 mg × 3 days) before anti-CD19/20 CAR T cells infusion. The dose is 1E5 ~ 2E6 anti-CD19/20-CAR positive T cells. The cells infusion process may last for 30 min

Primary outcome measures

Outcome measure	Measure description	Time frame
Occurrence of study-related adverse events	Defined as ≥Grade 3 signs/symptoms, laboratory toxicities, and clinical events that are possibly, likely, or definitely related to study treatment	From baseline to 12 months after

Secondary outcome measures

Outcome measure	Measure description	Time frame
Annualized relapse rate (ARR) of NMOSD Attacks	Compare annualized relapse rate before and 1 year after initial CAR-T administration	Baseline, 12 months
Change in expanded disability status scale (EDDS) score	The EDSS is an ordinal clinical rating scale ranging from 0 (normal neurologic examination) to 10 (death) in half-point increments	Baseline, 12 months
Change in best corrected visual acuity (Log MAR)	Visual acuity assessment through Snellen's test chart	Baseline, 12 months
Change in peripapillary retinal nerve fiber layer (pRNFL)	Compared pRNFL before and 1 year after initial CAR-T administration	Baseline, 12 months

Change in macular ganglion cell-inner plexiform layers (mGCIPL)	Compared mGCIPL before and 1 year after initial CAR-T administration	Baseline, 12 months
Change in flash visual evoked potential (FVEP)	Compared FVEP before and 1 year after initial CAR-T administration	Baseline, 12 months

Other outcome measures

Outcome measure	Measure description	Time frame
In vivo existence of tanCART19/20	RT-PCR (reverse transcription polymerase chain reaction) Analysis of whole blood will be used to detect and quantify survival of tanCART-19/20 TCR (T-cell receptor) zeta:CD137 and TCR zeta cells over time	Baseline, 12 months
Determination of serum immunoglobulins	Compare serum IgG level before and 1 year after initial CAR-T administration	Baseline, 12 months
Determination of serum AQP4 antibodies	Compare serum AQP4-ab titers before and 1 year after initial CAR-T administration	Baseline, 12 months
Determination of serum cytokines	Compare serum cytokines before and 1 year after initial CAR-T administration	Baseline, 12 months
Counts of peripheral blood B cell subsets	Compare peripheral blood B cells before and 1 year after initial CAR-T administration	Baseline, 12 months

Sponsor

Chinese PLA General Hospital

Collaborators

No information provided

Investigators

- Study Director: Quangang Xu, PhD, Chinese PLA General Hospital
- Principal Investigator: Huanfen Zhou, PhD, Chinese PLA General Hospital

General Publications

No publications available

Chapter 6
Terminated Ocular Stem Cell Studies

United States

First subject over 5 years after the transplant. The second subject is not in compliance with follow ups. We do not plan to enroll more subjects.

Hematopoietic Stem Cell Transplantation in Autoimmune-Related Retinopathy (ARRON)

ClinicalTrials.gov ID NCT00278486

Sponsor Richard Burt, MD
Information provided by Richard Burt, MD, Northwestern University (Responsible Party)
Last Update Posted 2013-04-08

Brief Summary
ARRON is a disease believed to be due to immune cells, cells which normally protect the body, but are now attacking the tissue in the retina and/or optic nerve. In addition, the disease may affect the nerves in the ear or other parts of the body. The affected nerves fail to respond, or respond only weakly, to stimuli causing numbing, tingling, pain, and progressive muscle weakness. If the nerves to the ear are affected, reduced hearing or deafness may result. The likelihood of progression of your disease is high. This study is designed to examine whether treating patients with high-dose cyclophosphamide and rabbit ATG (drugs which reduce the function of the immune system) followed by return of previously collected blood stem cells will stop the progression of ARRON syndrome. Stem cells are undeveloped cells that have the capacity to grow into mature blood cells, which normally circulate in the

J. N. Weiss, *Ocular Stem Cell Studies*,
https://doi.org/10.1007/978-3-031-58557-9_6

blood stream. The purpose of the cyclophosphamide and rabbit ATG is to destroy the cells in the immune system which are thought to be causing this disease. The purpose of the stem cell infusion is to restore the body's blood production, which will be severely impaired by the high-dose chemotherapy and to produce a normal immune system that will no longer attack the body.

Official Title
Immune Ablation and Hematopoietic Stem Cell Transplantation in Patients With Autoimmune-Related Retinopathy and Optic Neuropathy (ARRON) Syndrome (Not Associated With Cancer)

Conditions
Retinal Disease

Study Type
Interventional

Enrollment (Actual)
2

Intervention/Treatment
Biological: Hematopoietic stem cell transplantation.

Phase
Phase 1

Other Study ID Numbers
DI ARRON 04

Study Start
2004–08

Primary Completion (Actual)
2012–04

Study Completion (Actual)
2012–04

United States

Chicago, Illinois, United States, 60,611
Northwestern University, Feinberg School of Medicine

Eligibility Criteria
Ages Eligible for Study
18 to 60 years (Adult)

Accepts Healthy Volunteer
No

Sexes Eligible for Study
All

Inclusion Criteria

- Age between 18–60.
- Diagnosis of ARRON syndrome. Diagnostic criteria described below.
  Unexplained visual loss over weeks to months. The visual loss includes both visual acuity and field loss define as follows:

  - Visual acuity: 20/40 or less
    OR
  - Visual field: perimetric mean deviation -5b

      Positive antibody to retina or optic nerve.
      OR
      A response to immunosuppressive drugs or immune modulators (response is defined by improvement of vision or decrease the rate of decline of visual loss).
  - Absence of malignancy {negative physical examination, gastrointestinal endoscopies, mammography and gynecologic examination (for female), and serum PSA measurement (for male) within a year}.
  - Negative MRI of brain.

- The patient has failed at least 3 months of corticosteroids (prednisone 0.5 mg/kg to start), IVIG and at least one other immunosuppressive drug such as methotrexate, Imuran, cyclosporine, etc. Failure is defined by decline of visual acuity (by standard Snellen acuity clinical testing) or visual field (by Humphrey Automated Machine with the 30-2 program or using Kinetic Visual Fields on the Goldman Perimeter).

Exclusion Criteria

- Absence of light perception lasting more than 6 months.
- Any illness that in the opinion of the investigators would jeopardize the ability of the patient to tolerate aggressive chemotherapy.
- Prior history of malignancy except localized basal cell, squamous skin cancer, or carcinoma in situ of the cervix. Other malignancies for which the patient is judged to be cured, such as head and neck cancer, or breast cancer will be considered on an individual basis.
- Positive pregnancy test.
- Inability or unwillingness to pursue effective means of birth control. Effective birth control is defined as (1) refraining from all acts of vaginal intercourse (ABSTINENCE); (2) consistent use of birth control pills; (3) injectable birth control methods (Depo-provera, Norplant); (4) tubal sterilization or male partner who has undergone vasectomy; (5) placement of an IUD (intrauterine device); or (6) use, with every act of intercourse, of diaphragm with contraceptive jelly and/or condoms with contraceptive foam.
- Failure to willingly accept or comprehend irreversible sterility as a side effect of therapy.
- FEV1/FVC <60% of predicted after bronchodilator therapy (if necessary).

- DLCO <50% of predicted.
- Active ischemic heart disease and/or those who have had a myocardial infarction within 6 months.
- Resting LVEF <40%.
- Bilirubin >2.0 mg/dl.
- Serum creatinine >2.0 mg/dl.
- Known hypersensitivity to mouse, rabbit, or E. coli-derived proteins, or to iron compounds/medications.
- Presence of metallic objects implanted in the body that would preclude the ability of the patient to safely have MRI exams.
- Diagnosis of primary progressive MS.
- Platelet count <100.00/ul.
- Psychiatric illness, mental deficiency or cognitive dysfunction making compliance with treatment or informed consent impossible.
- Active infection except asymptomatic bacteriuria.

Design Details
Primary Purpose: Treatment
Allocation: Non-Randomized
Interventional Model: Single Group Assignment.
Masking: None (Open Label)

Arms and interventions

Participant group/arm	Intervention/treatment
Experimental: stem cell transplantation	Biological: Hematopoietic stem cell transplantation • Autologous hematopoietic stem cell transplantation will be performed

Primary outcome measures

Outcome measure	Measure description	Time frame
Standard Snellen acuity clinical testing and improvement visual fields is done by using Humphrey Automated Machine with 30-2 program or using Kinetic Visual Fields on the Goldman Perimeter)		5 years after transplant

Sponsor
Richard Burt, MD

Collaborators
No information provided

Investigators
- Principal Investigator: Richard Burt, MD, Northwestern University

General Publications
No publications available

Terminated

Reason: Based on a business decision unrelated to any safety concerns.

Long-Term Follow-up Safety Study of Human Central Nervous System Stem Cells in Subjects With Geographic Atrophy of Age-Related Macular Degeneration

ClinicalTrials.gov ID NCT02137915

Sponsor StemCells, Inc.
Information provided by StemCells, Inc. (Responsible Party)
Last Update Posted 2016-06-02

Brief Summary
This study is being done to determine the long-term safety and possible benefits of transplanted Human Central Nervous System Stem Cells (HuCNS-SC) for patients with Geographic Atrophy of Age-Related Macular Degeneration. This long-term follow-up study is limited to those individuals who received a transplant of HuCNS-SC cells into one of their eyes as part of the CL-N01-AMD study. No additional study product will be given in this 4-year long-term follow-up study.

Official Title
Long-Term Follow-up Study of the Phase I/II Safety and Preliminary Efficacy of Human Central Nervous System Stem Cells (HuCNS-SC) Subretinal Transplantation in Subjects With Geographic Atrophy of Age-Related Macular Degeneration

Conditions
Age-Related Macular Degeneration
Geographic Atrophy

Study Type
Observational

Enrollment (Actual)
8

Intervention/Treatment
Biological: Human Central Nervous System Stem Cells

Other Study ID Numbers
CL-N02-AMD

Study Start
2014-04

Primary Completion (Actual)
2016-05

Study Completion (Actual)
2016-05

United States

Dallas, Texas, United States, 75231
Retina Foundation of the Southwest

Eligibility Criteria
Ages Eligible For Study
51 years and older (Adult, Older Adult)

Accepts Healthy Volunteer
No

Sexes Eligible for Study
All

Sampling Method
Non-Probability Sample

Study Population
All subjects with age-related macular degeneration (AMD) with GA who received transplantation of HuCNS-SC in Study CL-N01-AMD.

Description

Inclusion Criteria
- Must have completed Study CL-N01-AMD
- Must be able to provide written informed consent prior to any study-related procedures
- Agree to comply in good faith with all conditions of the study and to attend all required study visits

Exclusion Criteria
- Inability to comply with study procedures or visits
- Since enrolling in Study CL-N01-AMD, have entered, or are about to enter another investigational study that, in the opinion of the Principal Investigator (PI), might confound study interpretation.
- Received off-study immunosuppressive agents in the lead-in study or are receiving any immunosuppressive agents since completing the lead-in study.

Design Details
Observational Model: Case-Only
Time Perspective: Prospective

Cohorts and interventions

Intervention/treatment
Biological: Human Central Nervous System Stem Cells • Other Names: – HuCNS-SC

Primary outcome measures

Outcome measure	Measure description	Time frame
Frequency and types of serious adverse events		At each of 11 visits over a period of 48 months

Secondary outcome measures

Outcome measure	Measure description	Time frame
Frequency and types of adverse events		At each of 11 visits over a period of 48 months

Sponsor
StemCells, Inc.

Collaborators
No information provided.

Investigators
- Principal Investigator: David Birch, Ph.D., Principal Investigator

General Publications
No publications available

Iran

Terminated

Reason: No improvement obtained in three cases. Retinal detachment observed in two cases

Stem Cell Transplantation in Ocular Lesions of Behcet's Disease

ClinicalTrials.gov ID NCT00550498

Sponsor Tehran University of Medical Sciences
Information provided by Tehran University of Medical Sciences (Responsible Party)
Last Update Posted 2015-11-10

Brief Summary
The purpose of this study is to find if autologous stem cell transplantation can stop the progression of intractable eye lesions of Behcet's Disease or even to improve it.

Detailed Description
To test in a pilot study the effect of autologous stem cell transplantation in stopping the progression of retinal damage of eye lesions in Behcet's Disease by repairing the vascular damage, and possibly to improve the lesions.

Official Title
Phase 1 Study of Bone Marrow-Derived Stem Cell in the Treatment of Ocular Lesions of Behcet's Disease

Conditions
Retinitis
Behcet's Syndrome

Intervention/Treatment
Biological: Autologous Stem Cell Transplantation

Study Start
2007-12

Primary Completion (Actual)
2012-07

Study Completion (Actual)
2013-07

Enrollment (Actual)
5

Study Type
Interventional

Phase
Phase 1

Other Study ID Numbers
3086

Iran, Islamic Republic of

Tehran, Iran, Islamic Republic of, 14114
Rheumatology Research Center, Behcet's Disease Unit (Shariati Hospital)

Ages Eligible for Study
16 years and older (Child, Adult, Older Adult)

Sexes Eligible For Study
All

Accepts Healthy Volunteers
No

Eligibility Criteria
Inclusion Criteria

- Behcet's Disease.
- Fulfilling the International Criteria for Behcet's Disease (ICBD)
- Ocular lesions with retinal vasculitis
- Resistant to at least 3 months of combination therapy with pulse cyclophosphamide (1 g/monthly), azathioprine (3 mg/kg/daily), and prednisolone 0.5 mg/kg/daily
- Visual acuity not less than 4 m finger count
- Retinal edema confirmed by color photography and OCT

Exclusion Criteria

- No Vision
- Active posterior uveitis
- Fundus not visible

Ages Eligible for Study
16 years and older (Child, Adult, Older Adult)

Sexes Eligible for Study
All

Accepts Healthy Volunteers
No

Design Details
Primary Purpose: Treatment
Allocation: N/A
Interventional Model: Single Group Assignment
Masking: None (Open Label)

Arms and interventions

Participant group/arm	Intervention/treatment
Experimental: A Behcet's Disease with ocular lesions.	Biological: Autologous Stem Cell Transplantation • Autologous bone marrow aspiration (20 ml) from iliac crest. Separation of mononuclear cells using ficoll hypaque and culture in 10% fetal bovine serum and Dulbecco modified Eagles medium. After confluent stage, detachment of cells with trypsin/EDTA, and subculture. Repeated passages until obtainment of required cell number. Confluent cells of last passage are washed with tyrode solution and incubated with M199 for 60 min. Cells are separated using trypsin/EDTA and washed 3 times with M199 and 1% HSA. Precipitate will be diluted with heparinized M199 to make solution with 6 × 106 cells/ml. Sample will be tested for viability and detection of CD45, CD34, CD90, CD44, CD13, CD105, and CD166 before injection. The number of cells for injection will be between 3-five million cells, in a maximum volume of 0.3 ml. Cells will be injected in vitreal at 3.5 mm of limbus with a 26 gauge insulin needle

Primary outcome measures

Outcome measure	Measure description	Time frame
1. Visual acuity by ETDRS (Early Treatment of Diabetic Retinopathy Study) 2. Retinal state and retinal edema by Color Photography, Multifocal Electroretinography, and OCT (Optic Coherent Tomography)		12 months

Sponsor
Tehran University of Medical Sciences

Collaborators
No information provided.

Investigators
- Study Chair: Fereydoun Davatchi, MD, Rheumatology Research Center, Medical Sciences/University of Tehran
- Principal Investigator: Hormoz Shams, MD, Rheumatology Research Center, Medical Sciences/University of Tehran
- Principal Investigator: Behrouz Nikbin, MD, Department of Immunology, Medical Sciences/University of Tehran

General Publications

- Otani A, Dorrell MI, Kinder K, Moreno SK, Nusinowitz S, Banin E, Heckenlively J, Friedlander M. Rescue of retinal degeneration by intravitreally injected adult bone marrow-derived lineage-negative hematopoietic stem cells. J Clin Invest. 2004 Sep;114(6):765–74. https://doi.org/10.1172/JCI21686.
- Kicic A, Shen WY, Wilson AS, Constable IJ, Robertson T, Rakoczy PE. Differentiation of marrow stromal cells into photoreceptors in the rat eye. J Neurosci. 2003 Aug 27;23(21):7742–9. https://doi.org/10.1523/JNEUROSCI.23-21-07742.2003.
- Jiang Y, Jahagirdar BN, Reinhardt RL, Schwartz RE, Keene CD, Ortiz-Gonzalez XR, Reyes M, Lenvik T, Lund T, Blackstad M, Du J, Aldrich S, Lisberg A, Low WC, Largaespada DA, Verfaillie CM. Pluripotency of mesenchymal stem cells derived from adult marrow. Nature. 2002 Jul 4;418(6893):41–9. https://doi.org/10.1038/nature00870. Epub 2002 Jun 20. Erratum In: Nature. 2007 Jun 14;447(7146):879–80.

Chapter 7
Suspended Ocular Stem Cell Studies

Transplantation of Cultivated Limbal Epithelium on Amniotic Membrane for Limbal Stem Cell Deficiency

ClinicalTrials.gov ID NCT00348114

Sponsor: Singapore National Eye Centre
Information provided by Singapore National Eye Centre
Last Update Posted 2006-07-04

Brief Summary
To evaluate the effectiveness of autologous (tissue from fellow eye) transplantation of cultivated limbal epithelium on amniotic membrane to treat severe surface irregularity and scarring of the corneal surface.

Detailed Description
The corneal epithelium is under constant cell-turnover, and it has been shown that the limbus is the ultimate source of epithelial renewal. Significant damage to limbal cells causes a disease state called limbal stem cell deficiency (LSCD), characterized by different extents of conjunctival overgrowth onto cornea, vascularization, chronic inflammation, and poor epithelial integrity.

In unilateral LSCD, the damaged corneal surface may be reconstructed using two large segments of healthy limbal tissue (of approximately 6–8 mm by 1 mm) from the fellow eye in the procedure of conventional limbal transplantation. This may however compromise ocular surface integrity of the healthy eye depending on the amount of donated limbus. In more recent years, it has been possible to "save" limbal tissue by obtaining only a very small limbal biopsy (2 mm by 1 mm) from the fellow healthy eye which is "expanded" into a viable sheet of limbal cells via laboratory cultivation. This cultivated sheet of cells is transplanted to treat the damaged ocular surface.

J. N. Weiss, *Ocular Stem Cell Studies*,
https://doi.org/10.1007/978-3-031-58557-9_7

This study is a prospective nonrandomized trial which evaluates the efficacy of transplantation of autologous ex vivo expanded limbal epithelium on intact amniotic membrane for unilateral total limbal stem cell deficiency.

Official Title
Transplantation of Ex-Vivo Expanded Human Limbal Epithelial Stem Cells (LSC) on Amniotic Membrane (AM) for Limbal Stem Cell Deficiency (LSCD).

Conditions
Ocular Surface Disease

Study Type
Interventional

Enrollment
8

Intervention/Treatment
Procedure: Transplantation of cultivated limbal epithelium

Phase
Phase 2

Other Study ID Numbers
R260/05/2002

Study Start
2002–05

Primary Completion
Study Completion
2006–05

Eligibility Criteria
Ages Eligible for Study
0 years and older (Child, Adult, Older Adult)

Accepts Healthy Volunteer
No

Sexes Eligible For Study
All

Description

Inclusion Criteria
- Unilateral total limbal stem cell deficiency with normal fellow eye, or minimally damaged fellow eye (less than 1/3 limbus affected)

Exclusion Criteria
- Systemic disease affecting both eyes such as Stevens–Johnson syndrome

Design Details
Primary Purpose: Treatment
Allocation: Non-Randomized
Interventional Model: Single Group Assignment
Masking: None (Open Label)

Arms and interventions

Intervention/treatment
Procedure: Transplantation of cultivated limbal epithelium

Primary outcome measures

Outcome measure	Measure description	Time frame
Snellen visual acuity		
Corneal epithelial integrity and stability		

Secondary outcome measures

Outcome measure	Measure description	Time frame
Extent of retarding recurrent neovascularisation		

Sponsor
Singapore National Eye Centre

Collaborators
Singapore Eye Research Institute

Investigators
- Principal Investigator: Seng-Ei Ti, FRCS(Ed), Singapore National Eye Centre
- Study Director: Donald TH Tan, FRCS(G), Singapore Eye Research Institute and Singapore National Eye Centre

General Publications

- Ti SE, Anderson D, Touhami A, Kim C, Tseng SC. Factors affecting outcome following transplantation of ex vivo expanded limbal epithelium on amniotic membrane for total limbal deficiency in rabbits. Invest Ophthalmol Vis Sci. 2002 Aug;43(8):2584-92.
- Tsai RJ, Li LM, Chen JK. Reconstruction of damaged corneas by transplantation of autologous limbal epithelial cells. N Engl J Med. 2000 Jul 13;343(2):86-93. https://doi.org/10.1056/NEJM200007133430202.

Chapter 8
Analysis

As of August 7, 2023, the clinicaltrials.gov website listed 461,430 studies being conducted in all 50 U.S. states and in 221 other countries. 12,882 or 2.8% are listed as utilizing stem cells, and of these, 206 or .04% involve the eyes. This number is actually inflated as many of the included studies were miscategorized.

However, if only Active/Recruiting studies are reviewed, there are 22 studies worldwide. Fourteen studies treat retinal and optic nerve diseases, and 8 treat anterior segment conditions. Twenty studies are corporate, university, or hospital supported. Two studies have no support and are patient funded (Tables 8.1 and 8.2).

J. N. Weiss, *Ocular Stem Cell Studies*,
https://doi.org/10.1007/978-3-031-58557-9_8

Table 8.1 Retina/optic nerve active recruiting studies

	UNITED STATES	EUROPE	EASTERN EUROPE	ASIA	MIDDLE EAST	CARIBBEAN/SOUTH AMERICA
RETINA/OPTIC NERVE STUDIES	7	1	1	3	1	1
CORPORATE	2			2		
UNIVERSITY/HOSPITAL/NEI	4	1	1	1	1	
NO SUPPORT	1					1
BONE MARROW	2					
IPSC	2			1		
ADIPOSE						
PROPRIETARY	3		1	1		
UNBILICAL				1		1
EMBRYONIC		1				
WHARTONS					1	
UNKNOWN						

Table 8.2 Anterior segment/active recruiting studies

	UNITED STATES	CANADA	EUROPE	ASIA	MIDDLE EAST	CARIBBEAN/SOUTH AMERICA
ANTERIOR SEGMENT STUDIES	2	1	2	3	0	0
NONPROPRIETARY	2					
PROPRIETARY						
CORPORATE						
UNIVERSITY/HOSPITAL	2	1	2	3		
OTHER SUPPORT						
NO SUPPORT						
BONE MARROW	1					
IPSC						
ADIPOSE						
PROPRIETARY				1		
UNBILICAL				1		
EMBRYONIC						
WHARTONS						
CULTURED LIMBAL CELLS/EPITHELIAL CELLS/ORAL MUCOUSA	1	1	2			
UNKNOWN				1		

Chapter 9
Comments

Clinicaltrials.gov Disclaimer

The U.S. government does not review or approve the safety and science of all studies listed on this website. The study sponsor or investigator submits information about their study to ClinicalTrials.gov and is responsible for the safety, science, and accuracy of any study they list.

- ClinicalTrials.gov is a website and online database of clinical research studies and their results. The National Library of Medicine (NLM) maintains the website. A broad range of public and private organizations around the world sponsor (oversee) and fund (pay the costs) studies listed in the ClinicalTrials.gov database. U.S. government agencies do not oversee or fund all the studies listed on ClinicalTrials.gov.
- Study sponsors or investigators provide the study information on ClinicalTrials.gov. The study sponsor or investigator is responsible for ensuring that their studies follow all applicable laws and regulations and for submitting complete, accurate, and up-to-date information about their study. NLM staff only reviews study information for apparent errors, deficiencies, or inconsistencies.
- Choosing to join a study is an important personal decision. Before you join a study, learn about the possible risks and benefits and discuss all options with your health care professional. To learn more about taking part in clinical research studies, read Learn About Studies, which includes questions to ask the study staff before deciding to join.

J. N. Weiss, *Ocular Stem Cell Studies*,
https://doi.org/10.1007/978-3-031-58557-9_9

What Are Stem Cells?

Stem cells are cells that can reproduce and have the potential to change or differentiate into another type of cell. At the present time, there are 3 main categories of stem cells.

Embryonic cells are taken or produced from very early fertilized embryos. They can differentiate into any type of cell, but this can be disadvantageous, as they can form tumors. Much of the early stem cell research was performed using this type of cell. Because of the source of their derivation, there has been much religious and moral controversy.

Federal funding was not allowed for embryonic stem cell research until President George W. Bush changed the policy. He allowed federal funding for stem cell research but only on those stem cell lines that were created before August 9, 2001, because those lines already existed. Funding was prohibited for any new stem cell lines. Unfortunately, only a few of these stem cell lines were useful for research.

However, New Jersey and California (2004), Connecticut and Illinois (2005), Maryland and Missouri (2006), and New York and Iowa (2007) eased restrictions and/or created funding opportunities for stem cell research. Other states enacted varying restrictions. In 2009, President Obama changed federal policy which allowed further research on new stem cell lines.

Is it any wonder that in an unstable political environment, where politicians are making scientific decisions, that researchers have left the U.S. to continue their work in other countries and U.S. companies refuse to invest in stem cell research?

Adult stem cells are found in cord blood, and different organs, including bone marrow, and fat or adipose tissue. The term "adult" may be confusing because they are also found in fetal tissue. These cells are further developed than embryonic cells. The cell is considered adult, not the source. Adult stem cells, unlike embryonic, generally form cells in their own tissue type, making the chance of tumor formation unlikely. They also provide many types of growth factors that may stimulate the existing tissue to heal itself.

Induced Pluripotent stem cells are differentiated, or older cells, that can be stimulated to return to an undifferentiated state. This would create an inexhaustible supply of cells but mutations in the cell line is a problem.

Stem cells were discovered in the 1980's, about the same time as seat belts were introduced in automobiles and computers had 64 meg hard drives. In the last 40 years there have been tremendous advances in automotive safety and in computers. Why has no similar progress been made in the field of stem cell research? That is the question we should be asking.

Who Pays for Research: Follow the Money

Unless you are independently wealthy and a philanthropist, someone must pay you to work. At the end of the day, you can have all the good intentions and empathy in the world, but you must eat, wear clothes, have a place to live, and if you have a family, support them as well. Nothing can be accomplished in science, or in life, without money. Rent, equipment, supplies, personnel, etc. must all be paid for.

A clinical trial, also called an interventional trial, is a type of research where participants receive a specific intervention that is dictated by a study protocol. The intervention may be a surgical procedure, drug, device, or change to the participants behavior, such as a diet. Clinical trials may compare a new medical or surgical approach to no intervention. There are various sources that provide funding for clinical trials.

The National Institutes of Health (NIH)

The NIH is the largest biomedical funding source in the U.S. There has been a 20% inflation adjusted decline in the NIH budget over the last decade. Approximately 15–20% of grant submissions receive funding and due to government budget sequestration, investigators receive only 90% of the approved funds. A 2007 U.S. government study found that university faculty members spend approximately 40% of their research time navigating bureaucracy in order to receive funding for their work.

A research grant is awarded to an individual at a university, but not to the same individual when they enter private practice, effectively removing an experienced researcher from government funding. NIH grants are typically for a 3 year period and end abruptly if not renewed. In addition, the researcher is unable to shift the funds to another project if the first project is not successful.

Pharmaceutical Companies

Pharmaceutical Companies are commercial enterprises organized to make money. Approximately 70% of clinical trials in medicine are company sponsored. Such funding may introduce bias, as the study design or interpretation is more likely to favor the drug under consideration with the ultimate goal, the attainment of FDA approval. This is understandable as the time to develop a successful drug is 10 to 15 years and the cost to achieve FDA approval has successively increased over the decades to more than 1.2 billion dollars at the present time. Ultimately, only 12% of studied drugs obtain FDA approval.

An orphan disease is defined as any disease affecting up to 200,000 individuals in the U.S. There are incentives offered to pharmaceutical companies through the Orphan Drug Act of 1983 to address the rare disease market. However orphan diseases with larger numbers of patients or those diseases for which there may be less price resistance regarding any successful treatment may draw development efforts in an unfair way.

Non-Profit Organizations

Like the NIH, non-profits use a panel of experts who decide the worthiness of a particular grant and determine the awarding of funding. As such, they suffer from the same problem as the NIH.

Thomas Kuhn has pointed out that breakthrough insights frequently stem from the intersection of disciplines, not from within the discipline. Grant reviewers are generally within the same discipline and don't recognize the new paradigm that may have come from the work of scientists in other fields. They remain committed to the old paradigm that has shaped their beliefs, which prompted the physicist, Max Planck, to say, "New scientific truth does not triumph by convincing its opponents and making them see the light, but rather because its opponents eventually die, and a new generation grows up that is familiar with it." This leads to the expression, "medicine is advanced one funeral at a time." Also like the NIH, non-profit organizations award grants to institutions, not to individual researchers in private practice.

Private Funded Research

This includes a broad pool of donors with a wide range of passions that may speed progress by investing in bold ideas, gambles, and risky projects. Decisions may be made quickly, not encumbered by large bureaucracies. A sense of urgency can be respected.

Critics argue that gifts privatize research and steer resources towards areas of personal interest. Supporters argue that it is the personal choice of the funder whether money is spent on personal goods or personal research. Personally funded projects have a vested interest in solving the chosen problem and unlike pharmaceutically funded research, are not subject to the profit motive or market forces.

Patient Funded Foundations

It is not uncommon for many patients with rare diseases or their families to set up support groups, particularly with the ready availability of social networking. In some cases, the only research foundations focused on particularly rare diseases have been initiated by patients and families affected by those diseases. There are over 7000 rare genetic diseases impacting 8–10% of the US population or about 25 million patients. When such a disease afflicts a high profile or financially successful individual, or their family member, that particular disease can receive disproportionate attention and support.

Patient Funded Research

In this model, the patient pays for participating in the clinical research. In terms of the research, this is similar to the Private Funded model discussed above. Less common conditions and risky proposals may be studied, unencumbered by bureaucracy, and coupled with the ability to make quick decisions.

Critics argue that desperate patients can be exploited and should not pay for clinical research. Clearly, people with resources decide how to use their resources. Not allowing patients to pay for research would restrict their ability to use their own resources as they see fit.

Unlike other types of funding, patient funded research is subject to market demands. If a study is too expensive, there will be less participants, unlike the situation where the funding is provided for the patient to participate in the research. The self-funding patient will be more inclined to demand information thus providing a more thought out informed consent than in a study where free care is provided, as they want to receive their "money's worth." One can argue that free medical care may be considered an exploitative inducement to participate in research.

Patient funded studies are generally for conditions not being investigated by NIH or pharmaceutical studies and tend towards novel approaches for less prevalent medical problems.

Patient funded research has been criticized for, by its nature, it eliminates the participation of individuals who cannot afford to participate. One can make the same criticism for the other funding sources as unless the funded study pays the patient for participation, including travel costs, individuals may be unable to take the required time off from work, nor have the financial and personal support structure in place to engage in a study.

Since the purpose of a research study is to determine whether the clinical intervention is of benefit, and such interventions may carry some medical risk, one could argue that aligning financial cost with participation creates a higher level of patient engagement, risk benefit assessment and therefore informed consent. In addition, the ultimate benefits of basic and clinical research accrue to all individuals with a particular disease regardless of financial wherewithal.

Does profit taint research? All human activity creates bias, otherwise there would be no activity. Unless the research is self-funded, researchers are always paid by some funding source, otherwise they could not perform the research. Frequently, both NIH and pharmaceutical funding is proportional to the number of subjects the researcher enrolls in the study, which some may consider to be unethical. In this respect, the patient funded study differs little from other types of studies. As pointed out previously, it is in the researcher's interest to keep the patient cost at a reasonable level in order to encourage patient entry.

The patient funded study, that has a detailed protocol, has undergone Institutional Review Board review, is listed with ClinicalTrials.gov (NIH), is collecting data and is publishing results, whether positive or negative, is an important form of research allowing the testing of ideas that would otherwise languish due to a lack of funding.

All human endeavors require work and effort and therefore have financial costs. There is a range of how removed the patients are from those costs depending on the research payment methodology.

In the case of publicly funded clinical research, the patient may have low awareness of the cost of the clinical research as the funds are taken from the general tax base of whose allocation individuals have minimal awareness. In publicly funded research the cost of an individual's participation is spread out over millions of people and therefore the percentage of cost that is the responsibility of an individual patient is low. This low awareness of source and percentage cost of participation removes these particular financial considerations from the informed consent process.

On the other end of the spectrum, patient funded research causes high awareness of the cost of that individual's participation and causes a high percentage of the cost to be the responsibility of the participating patient. This high awareness enters the informed consent process as an additional consideration as patients weigh the potential personal benefit of participation. This causes a pause in this process, even for the wealthiest individuals.

Pharmaceutical clinical research would fall in-between the two, with the actual costs being deferred to a later time, but still assumed by the individual patient needing care depending on the health insurance methodology.

In considering the different methods of funding of clinical research, the overall benefit of properly done clinical research, from whichever funding source, should be kept in mind. All clinical research and patient care create costs which ultimately accrue to individuals in the society. It is only a matter of how directly those costs are felt by individuals.

Patients are not uncomfortable with the thought of having to pay for an experimental treatment for a condition in which they have been repeatedly told there is no treatment and no hope for improvement. Patients are fully aware that they must pay for treatments or surgeries not covered by their insurance and know that experimental treatments are not paid for by insurance. Interestingly, a common criticism from physicians to their patients is that if they have to pay, it is unethical and immoral, because physicians are familiar with corporate or government sponsored studies in which the care is "free."

Is it ethical or moral to insist that a patient with an "untreatable" condition return to that same physician for follow-up and extensive testing for which the patient and/or the insurance company is charged, when the physician has no treatment now, or in the foreseeable future, nor any intention of looking for one? Saying that, "they are working on it," when no one is, is doing the patient a disservice. The patient knows that they are losing their vision, what value is it to them to pay for the physicians "observation," similar to watching a house burn down without intervening?"

Having participated in many research studies funded by the government or drug companies the physician always gets paid, and in the case of the drug companies, very lucratively. The physician never works for free. If he is an employee of a hospital or a university, they pay him, in additional to money he receives from performing a study and/or being a "consultant," (the "consultant" title and payment is used to curry physician favor).

Physicians in private practice are not eligible to obtain government grants. The grants are awarded to physicians at universities. The people making the grant decisions are academicians and they support their own. 1 day the people they award the grant to may sit in judgement of their grant. Everyone knows one another and one hand washes the other. A non-academic physician is closed to this funding source, and a wealth of knowledge and intelligence is lost as a result.

One would think that a private foundation would be eager to fund promising research but that is not the case. The same academicians sit on the board of the foundations and, as above, fund their own friends and pet projects. In some cases, the supposed independent non-profit foundation is industry funded, or has ties or investments with industry and is not interested in proposals against their own self-interest.

When trying to obtain FDA approval for their drug, studies require "pure" conditions in generally young patients, excluding longstanding visual loss in older patients who may have multiple medical problems. Isn't this latter population entitled to a chance? If a study only enrolls patients with "untreatable" conditions, in which the natural history of the condition is such that no one has ever improved, or is a progressive condition in which everyone has lost their vision, and the patient undergoes an "experimental" treatment and their vision improves, is that chance? And you can repeat the treatment and most patients gain vision. Would a double-blind study be truly necessary? And knowing this, how would you feel if your wife, or child were in the control, untreated group? Does the sick patient benefit when limited dollars are spent on procedure and satisfying a checklist, and not on treatment?

Top tier publications and newspapers recognize these facts. Home town newspapers report the success of their neighbor who benefited from treatment. It is the second, and the third and fourth tier publications that feel the need to demonize stem cells with negative and sensational reporting in order to sell their publications. An inaccurate, disingenuous article frightens patients away from a treatment that might potentially help them and extinguishes their only glimmer of hope.

It is very easy to criticize someone else. Theodore Roosevelt said it best in "The Man in the Ring":

"It is not the critic who counts; not the man who points out how the strong man stumbles, or where the doer of deeds could have done them better. The credit belongs to the man who is actually in the arena, whose face is marred by dust and sweat and blood; who strives valiantly; who errs, who comes up short again and again, because there is no effort without error and shortcoming; but who does actually strive to do the deeds; who knows great enthusiasms, the great devotions; who spends himself in a worthy cause; who at the best knows in the end the triumph of high achievement, and who at the worst, if he fails, at least fails while daring greatly, so that his place shall never be with those cold and timid souls who neither know victory nor defeat."

Academics is not the same as it used to be. Financial pressures have changed it into a moneymaking endeavor. You can make a career out of criticizing someone. You will become the "go-to" person for people who agree with you. Prior to cable news, everyone listened to the same news from one of a few network television stations. Now with cable television, you can pick the TV station that agrees with your opinions and never hear anything that disagrees with you. Objectivity has been replaced with spin. One expects a true academician to do independent research, not copy from a newspaper article, and claim the information as his own. This has become his moment in the sun, and the person to look to if you are against stem cells, because like the cable news analogy, he will agree with you, whether you are right or wrong.

And why would you be against stem cell research? Well, one reason is if you have ten million dollars in grants, and have accomplished nothing, then it would be your imperative to criticize those that have accomplished something with little or no funding by saying their work is suspect and they didn't play by the rules. But when patients are suffering, and the rules are made by those with vested interest in their own turf, they need to be changed or ignored.

I have had physicians who refused to speak to me for many years, suddenly, out of the blue telephone me. After the obligatory "how are you doing" they get to the point that they, or a family member, has a problem and can I help them? The shoe is on the other foot. It's easy to be righteous and make so-called "ethical" decisions for others when you have no "skin in the game." Everything changes when it affects you or someone you care about.

There are thousands of regulators and attorneys who make a living regulating and serving as consultants to those companies, universities, etc. who hire them to help them navigate the complicated regulatory process. This is their turf and they don't want anyone changing their world and affecting their income.

Academicians want to protect their turf as the keepers of the "holy truth." How can they be at the mecca of knowledge when someone in private practice is helping patients and they are not? They make a lot of money receiving government grants and money from drug companies and have a vested interest in protecting their source of income.

The Right-To-Try movement argues that if patients are sick or dying, they don't have the time to wait for such large-scale expensive studies, that may never happen. They may not be eligible for the study for whatever reason and aren't worried about possible side effects. When the end is inevitable, they want to be in control and try to help themselves. Shouldn't patients be allowed to control their own destiny? Aren't you allowed to purchase things based upon your own finances or health? It's your right to do what you wish. Life has risk. If someone has an automobile accident, would automobiles be banned? Trying to get well is a person's right.

Thoughts

Thomas Jefferson changed John Locke's "life, liberty, and property" to "life, liberty and the pursuit of happiness" which the Declaration of Independence guarantees as "natural" rights. The founding fathers felt that human beings have the natural intelligence to pursue happiness. It is self-evident that the pursuit of happiness would include the right to "get well." Yet the FDA has denied sick patients this "unalienable" right to choose treatments that they don't approve of.

The regulatory function of the FDA began in 1906 with the passage of the Pure Food and Drugs Act. There is no statutory language barring the FDA from regulating the practice of medicine though the agency has consistently asserted that it does not do so. It's actions however, demonstrate the opposite.

Stem cells were discovered 40 years ago; why have the promises of regenerative medicine not been realized? During World War 2 factories worked around the clock manufacturing war materiel. An entire Liberty ship was once built in 5 days. Yet, as the late actor Christopher Reeve, paralyzed by a horseback riding accident, pointed out, there is no sense of urgency in helping patients.

Geron Inc. developed a promising stem cell treatment for patients with spinal cord injury. They abandoned their efforts, not for scientific reasons, but as a business decision. ACT Inc. developed a stem cell treatment for several eye diseases. Their application to the FDA was more than 20,000 pages, the largest application the FDA has ever received. Did the time and expense of preparing that application benefit patients? No. ACT experienced consistent financial difficulties until the it was eventually sold to a Japanese company for their patents.

Business investment requires legislative stability. If the bar to perform stem cell work is high and constantly increasing in height, only a few, large U.S companies can risk the investment. Research will be performed in other countries and sick patients, with no other hope, will travel to those countries for treatment. The U.S. Passport affords its citizens the protection of the U.S. Government. Yet it is the absence of the protection of their rights to make decisions about their own healthcare that drives patients out of their own country.

We have the right, as citizens, to use tobacco products, drink alcohol, purchase non-FDA regulated vitamin and mineral supplements, drive an automobile, purchase firearms, engage in risky behavior and make poor social and financial decisions, yet the FDA, through its actions, restricts the ability of physicians to offer "unapproved" treatments and forces companies to spend hundreds of millions of dollars to enter the FDA system of regulation, when little of that funding will ultimately benefit the patient. The millions of dollars required to run the FDA gauntlet for approval restricts the development of new treatments and devices to only those with large markets with an ultimately expensive drug or treatment that warrants the investment. Cast to the wayside are inexpensive devices or drugs that benefit a small population.

Gutenberg's invention of the printing press removed the monopoly a small group had over knowledge and expanded it to the masses. The Internet has done the same. The sick patient is now aware that they have been lied to, that the phrase "they are working on it" that they have heard, in some cases, their entire lives, is untrue, when no one is working on it. In the past patients were cowed by the medical establishment's mystery and their own inability to obtain the most recent information. This has completely changed with the Internet and the ready availability of every article on every advancement being made.

Patients are incensed that progress is reported in basic science and animal studies but is not being advanced into the clinic. Establishment academicians and traditional physician leadership of medical conferences/journals are frightened of losing control of medical advancement and being relegated to the backwater of medical progress - hence their outcry for patients to let the traditional slow wheels of potential progress grind. They are not the ones with the "untreatable" medical conditions. Opinions change quickly when you or someone close to you, is affected.

Patients are now refusing to suffer while mainstream medicine diddles. Every patient who obtains treatment overseas and returns with positive news generates others who move forward. The internet and social media provide instant communication of good or bad results and link patients together in ways that were never possible before. The physician leadership of medical conferences, journals and universities struggles assiduously to remain in control of patient choice, but as a consequence of patient access to open-source knowledge, their model no longer works.

FDA's fundamental paradigms to control treatment access worked well during the decades when communication about disease and treatment was tightly controlled by physicians and government. Now that information is openly available there is no putting the genie back into the bottle.

The blind patient, the paralyzed patient, the patient dying from Lou Gehrig's Disease or Alzheimer's disease, have been betrayed by the trust they once had in physicians and the government, despite protestations to the contrary, that regulates their actions. To paraphrase Jefferson, the power to regulate is the power to obstruct. The phrase "the FDA stifles innovation" is frequently heard when what really should be said is that by stifling innovation, patient suffering is prolonged. Overregulation produces stagnation which hurts, not protects patients. The right to choose is a very powerful inherent right. Knowing that you tried, even if unsuccessful, is an important human emotion, more so in a sick or dying patient.

Clearly, autologous tissue is not a drug. Drugs are manufactured according to Good Manufacturing Practices (GMP). GMP are strict regulations that assure the manufacturing safety of the mass distribution of drugs. The patient taking the drug does not sign an "Informed Consent" when taking the medication. On the other hand, a patient undergoing a medical procedure must provide informed consent which consists of a thorough understanding of the risks and benefits of the proposed treatment.

There is no mass production or distribution and the risks and benefits of the procedure can only be assessed by the physician providing the therapy. The patient must then make the decision to undergo, or not undergo the procedure. This is clearly the practice of medicine. All surgical procedures have been developed in this fashion.

The medical profession self-corrects by education, licensure, publications and peer review. Beneficial procedures are adopted, and those without benefit, discarded. When the FDA began to regulate the use of autologous stem cells, organizations such as the American Red Cross, the American Society for Clinical Oncology, the Society for Assisted Reproductive Techniques, among others, protested, but to no avail.

By 2005 the FDA authority was expanded to make "more than minimally manipulated" allogeneic cells the same as prescription drugs. In 2006, under the guise of clarification, they further expanded their authority to include autologous cells and tissue. This was accomplished by changing one word, "into another human" was changed to "into a human" further expanding their authority into a wide variety of surgical procedures. Now, if a plastic surgeon uses the patient's own adipose tissue to perform a breast reconstruction, the FDA can claim that this is non-homologous use and the adipose tissue is now a drug.

Since stem cells by definition produce various tissue types, to characterize the placement of cells in another location is illogical and incorrect. Common sense has been lost. Unchecked power expands into vacant space without oversight.

Congress should must to protect patients, alleviate suffering and allow innovation. FDA interference in the physician development of new regenerative medical therapies must be removed. You can overprotect patients so much that nothing can get accomplished. That is why stem cell research has been obstructed for the last 40 years.

Proponents of increased FDA regulation cite the tragedy of thalidomide. Thalidomide was developed in the 1950's and sold as an over-the-counter mild sleeping pill that also reduced morning sickness, so it was popular with pregnant woman. Unfortunately, there was no animal testing of the drug in pregnancy. By 1960, doctors were concerned about the increasing side effects that were observed, including birth defects. The German company that manufactured the drug was unable to provide evidence to refute concerns.

Clinical drug trials were not necessary at the time but a dedicated FDA reviewer refused to approve the drug. The manufacturer still was able to import the drug to the U.S. and babies were born with birth defects. Worldwide drug laws changed after this tragedy. Thalidomide is still successfully used to treat leprosy and some AIDS-related conditions and its clinical analogues are being used to treat some types of cancers and immune disorders.

But what has happened since? At least 35 FDA approved prescription drugs have been removed from the market when it was discovered that these medications caused serious side-effects not known at the time of approval. The removed drugs had been on the market from 0–30 years with many complications including deaths reported.

How do we change things?

This is the bill I wrote that I would like to see Congress pass:

Title: Access to Innovation in Health Sciences.

Preamble:

WHEREAS the present cost of completing and implementing an application to the FDA is time consuming and expensive such that only large companies and institutions are able to absorb the cost;

WHEREAS individual physicians and researchers are unable to participate in the process, thus their innovative ideas are excluded from public utilization;

WHEREAS individual physicians and researchers, small companies, and institutions, may wish to perform proof of concept studies to determine whether investment in a larger study is feasible;

WHEREAS larger companies and institutions may wish to perform a small preliminary study to investigate whether further investment is warranted;

WHEREAS in order for the U.S. to remain competitive in the health care market all promising ideas should have the ability to be tested and scrutinized.

Section I

The FDA will authorize the creation of two new agencies; The Department of Innovative Therapies, and The Department of Innovative Devices. Applicants will apply to the appropriate agency depending on whether they are researching the use of a new therapy or a new device.

Section II

Studies that have been successfully reviewed and approved by an Institutional Review Board (IRB), and are registered with the NIH website, ClinicalTrials.gov, are eligible for submission to these agencies.

Section III

The application process will consist of the submission of:

1. The IRB approval letter.
2. A listing on ClinicalTrials.gov.
3. A copy of the Research Protocol.
4. A copy of the Patient's Informed Consent.
5. A one-page summary of the value of the proposed research.

Section IV

Each Department will have its own committee consisting of 3 physicians, an attorney, and a layman. The Committee Chairman will be a physician. Each Committee member will have an alternate if the Principal member is unavailable.

Section V

Applications will be approved by majority vote within 90 days of submission. Approved applications will be listed as: FDA Approved-Provisional.

Section VI

The approval will be for a term of 2 years which may be extended with subsequent applications.

Section VII

It is intended that the FDA Approved - Provisional designation will provide the researchers with liability protection.

Section VIII

The groups applying for this certification will pay a $250.00 fee for the application process.

Section IX

At the completion of the study a Final Report will be submitted to the Committee. If the therapy or device demonstrates treatment effectiveness with reasonable safety considering the nature and severity of the disease or impairment, the therapy or device will be deemed approved by the FDA.

Section X

It is anticipated that Medicare, Medicaid and private insurance companies will benefit from this work and should contribute to the Department's funding.

Section XI

Products approved through this mechanism are subject to price controls.

Section X

This bill shall go into effect ninety-one (91) days after passage.

Let's look at some of the other reasons stem cell research has been stymied over the last 40 years.

What is Bias?

Bias is an opinion not based on evidence or fact. It is the absence of an open mind and is related to prejudice, or forming an opinion before knowing the relevant facts and having an attitude resistant to rationalism. There are many types of bias.

Cognitive bias is an error in thinking where people create their own reality from their own perception. This can lead to irrationality, although in some cases may be adaptive allowing speedier decision making.

Anchoring is the propensity to use the first piece of information as an anchor in making decisions. The first price on a house or car serves as the basis for negotiation, even if it much higher than the item is worth.

Apophenia is the tendency to see patterns within random data. Seeing a pattern in the winning numbers in roulette is an example.

Attribution bias occurs when we assign motives to people's behavior which may not reflect reality.

Confirmation bias is the tendency to recall, search for, and interpret information that confirms one's own opinion while giving less attention to the information that contradicts it.

Framing is used by political parties to present facts in a manner in which their solution is the answer to the problem they presented. Cultural bias is a related phenomena.

Halo effect occurs when one's overall impression of company, brand, person, etc. influences their feelings about them. If you like one aspect of something, you will be positively predisposed to other things about it.

The Horn effect is the opposite of the Halo effect. In this case if a person holds a negative feeling towards something, the negative predisposition will translate to everything else about it.

Self-serving bias is when one own's self-esteem colors or distorts their perceptual processes such that accomplishments are credited to their own efforts, but failure is caused by outside elements.

Conflicts of Interest

A conflict of interest presents the risk that one's judgement or actions will be unduly influenced by another outside interest. They include: bribery, favoritism, insider trading, lobbying, sports fixing, shilling (such as a shell game), and regulatory and funding issues.

If an organization self-regulates, like a medical academy, a conflict of interest may be created as their committee members or the academy may derive secondary benefits from their actions. When a major portion of their funding is provided by pharmaceutical companies, and the same companies lucratively pay their physicians to run studies and serve as supposed "consultants" then the Academy has a vested interest in removing any threats to their income. It is a question of "following the money." The perception of impartiality is more important than the reality of impartiality.

The U.S. Federal Trade Commission issued an opinion of concern that the American Academy of Ophthalmology could use their supposed "ethics committee" to stifle innovation, but as a federal court recently ruled, the Academy doesn't have to follow their own rules, and they are above the law when attacking members. The Judge made an intellectually lazy decision by taking the path of least resistance as it would have been much more likely that a large organization with unlimited resources is more likely to appeal a negative decision than an individual. Unfortunately, by shirking responsibility, his decision negatively affects public health by preserving a monopoly which stifles any innovation, unless they financially benefit from it. To paraphrase the Roman philosopher Cicero, money trumps integrity.

A regulatory organization, intended to act in the public interest, may instead act in the interest of the groups it is charged with regulating. As an example, FDA regulators and attorneys may curry favor with pharmaceutical companies with the hope of future employment (which happens). In order to maintain an intellectual balance, they "over-regulate" other groups. Political and social influence may also affect regulatory independence.

Academics today is not the same as it once was due to increased financial pressures. This is reflected in published journal articles. Advances are few. Most studies are drug company sponsored, or a discussion of an exceedingly rare clinical condition. Why does someone become an academician today? Unless they genuinely want to advance the field and have an interest in seeing more complicated cases, the answer is ego. Unfortunately, if one is devoid of original ideas, they can advance their careers by attacking other people who have ideas. The best candidate to attack is one who is not in academics, and who cannot defend themselves because the journals are controlled by the academics, and they protect their own. The hallmark of a dishonest academic is the one who knows the facts, and lies about them, or who doesn't know the facts but rather than study them, confabulate ones based on his bias. And unethical reporters delight in the supposed expert who agrees with their premise that a great wrong is being done. You advance your career on the backs of others.

How can an academic explain and justify that the millions of dollars they received in research grants have not resulted in anything beneficial, other than paying the salaries of the physicians who write papers based upon a chart review, an esoteric condition, or in attacking a physician who truly developed something new?

Another person to watch out for is the supposed expert with a blog. They aren't interested in balanced reporting and, like the intellectually dishonest academic, only report items to bolster their animus to stem cells. Except, when the facts don't support their point of view, they lie. The dishonest reporter revels in this type of parasite.

Funding bias is the tendency of the research paper to support the interests of the financial sponsor. This is a well-known problem in medicine, especially in pharmaceutical company sponsored research.

Publication bias is also a well-known problem in medicine. When you submit an article to a journal, the Editor refers the article to anonymous reviewers that he selects. These reviewers may be friends, enemies, jealous, or competitors of the paper's authors. The submitted article may be "sat on" for months, and then rejected, inappropriate or personal attacks may be sent to the authors with the returned article, or the authors may be asked to make revisions to the article, and then once those revisions are made, continually request further revisions, so the article is never published. Controversial topics, such as stem cells, may be rejected outright, because they are new, and are not accepted by the "traditional" physicians.

Unfortunately, since the public views the publication in peer reviewed medical publications as the "gold-standard" in medicine, unaware of the problems with this system, truly avant-garde discoveries may not see the light of day. In addition, when work is attacked by the establishment who controls who gets published, further proof of their intellectual dishonesty is that they refuse to publish corrections, or let the attacked party answer the criticism, the result of which is that the inaccurate information remains in the public record.

Academic bias allows the researcher to shape their research and their opinions to conform with their own beliefs. In experimenter bias, the researcher, whether consciously or unconsciously, skews the experimental results. This is easy in human studies in which the recording of the patient's vision may be affected by bias.

Specifically, patients with poor vision may need additional time to focus on an object. If the examiner rushes the patient, the recorded visual acuity will be poorer than if the patient is given adequate time to see the eye chart.

Obviously, this problem isn't unique to stem cell work. Harold Ridley, the ophthalmologist who invented the intraocular lens was ridiculed to the point where he said he was sorry he invented it. The powers that be said that the people who first used laser to treat diabetic retinal disease were insane. Charles Kelman, who invented ultrasound to remove cataracts, Dr. Atkins and his diet, Barry J. Marshall and Robin Warren who discovered the bacterium Helicobacter pylori as the cause of ulcer disease, the list goes on and on of people ridiculed and almost destroyed because their ideas bucked the predominant traditional opinions. First, they ignore you, then they attack you, then the younger physicians say that they knew it all along.

Why Do People Resist Change?

Max Planck, the great German physicist, said: "An important scientific innovation rarely makes its way by gradually winning over and converting its opponents: What does happen is that the opponents gradually die out."

It is human nature that people will be more comfortable with what they already know, and that there is uncertainty in change. "Better the devil you know than the devil you don't know." It is about giving up something old and comfortable, for something new and possibly uncomfortable. People like longevity and tradition. Change is work, and people tend to resist work. Those people, who built their reputations on the past, will be defensive. They dread the perception that they were wrong and wish to maintain the status quo. They may wonder whether they can learn the new procedure, questioning their competence. More work is necessary, there is always a fear of failure. You may need the cooperation of past enemies, and resentments resurface. It is always easier to say "no" than "yes." Change for some people is very difficult. When they occupy a position of power, they have the ability to deny improvements to others.

People generally don't resist change when they believe it is in their best interest. It is easy to change if you win the lottery. In medicine, it is unfortunately common that physicians/researchers will take someone else work and claim it as their own, or if they can't, discredit it.

Doctors are like other people, they have egos (probably more so) and jealousy, greed and all the motives other people have, despite the Hippocratic Oath they take to put patients ahead of themselves.

You would expect a tax accountant to be current with the tax laws much as you would expect a lawyer to be current with the laws in his field. Yet a doctor can make the blanket statement that he never heard of any stem cell work, or say something else that is completely inaccurate. The patient will return home blind and without hope, and the doctor is unaffected.

To say that there are no treatments or no proof for something implies that you looked. Patients have read these studies on clinicaltrials.gov but the doctors, even after the patient has told them about it, refuse to look. Yet, when it affects someone in their family, as has happened, they become experts in this work and contact us.

Why Are We Obstructing Innovation?

The National Institutes of Health (NIH) was the major source of medical research funding and the largest medical research institute for decades. Over the last 30 years private industry now spends more than double the NIH on medical research. 85% of the new approved drugs have come from industry not government research.

Over the last decade, research into finding the mechanisms of biologic phenomena has made how something works seem more important than whether it does work. Many of the medical advances, like antibiotics, vaccines, x-rays, etc. were discovered by chance, not molecular analysis.

The driving force of scientists being recognized for their work drives them into elegant and novel solutions to problems when a simpler solution may be all that is necessary. The publication of positive findings in a prestigious journal affords the researcher promotion and tenure. The Editors of the prestigious journals exercise monopoly power over knowledge in terms of what they choose to print. Innovation requires the researcher to look across various fields, to form a team, and to think out of the box, an anathema to academicians who celebrate individuals like themselves.

As academic institutions use NIH funding to pay physicians salaries, and NIH funding is steadily decreasing, supposed physician researchers must spend more time treating income producing patients, and not perform research. Large research laboratories are underutilized. The chance of receiving a grant from the NIH is small, the duration of the grant award is shorter, and the NIH wants the work they fund to be successful, thus forcing the researcher to essentially perform the experiment before receiving the grant. The scientist must concentrate on short term results, long term projects have little chance of funding. Funding limitations preclude taking risks. Most of the U.S. ophthalmic research published today is funded by drug companies, whereas non-drug company funded scientific studies are being performed outside the U.S.

A basic problem of the entire system is peer review. Reviewers of a proposal base their recommendation of the knowledge in their field. They wish to deepen their own knowledge and reputation and can't or aren't interested in work outside their field. Yet breakthroughs generally occur at the intersection of disciplines. Researchers from other fields examine the problem from different points of view. Yet as past history has shown, things change in reaction to success, rarely to enable success.

I hope my efforts in presenting this volume will result in an increased awareness of the work presently being performed and a renewed interest in developing new ideas. We all become patients at some point in our lives.

Index

J. N. Weiss, *Ocular Stem Cell Studies*,
https://doi.org/10.1007/978-3-031-58557-9

MIX
Papier aus verantwortungsvollen Quellen
Paper from responsible sources
FSC® C105338

If you have any concerns about our products,
you can contact us on
ProductSafety@springernature.com

In case Publisher is established outside the EU,
the EU authorized representative is:
Springer Nature Customer Service Center GmbH
Europaplatz 3, 69115 Heidelberg, Germany

Printed by Libri Plureos GmbH
in Hamburg, Germany